SAT Math Success:
A Focused Approach

by

Richard Shedenhelm

CONTENTS

Introduction   v

1. Solving Simple Equations   1

2. Solving Quadratic Equations   7

3. Function Notation   21

4. Systems of Linear Equations   27

5. Complex Numbers   43

6. Geometry   55

7. Trigonometry   65

8. Randomized Problem Set 1   73

9. Randomized Problem Set 2   85

INTRODUCTION

Early last year, the College Board launched a new version of the SAT, posting 280 math practice problems consistent with the updated test. These problems showed up on four practice exams for the SAT and one practice exam for the PSAT. At Athens Academy, a local private college preparatory school, I used these problems to test the students I was teaching. The students in my class had a wide range of math abilities, but some of them were the top math students in their grade. Nevertheless, *all* of them had significant difficulties with the types of problems included in this book. Every student I have tutored individually also found these problems challenging. This experience gave me one criterion to include a topic, viz., the apparent difficulty of the skill. The second criterion was that these categories of math questions come up repeatedly in the practice exams.

I have tailored the depth of explanation and the quantity of repeated drill to match what I have found is needed by the students I have taught and tutored. For example, I have found that students need a great deal of drill in computing complex number problems correctly, but little drill in solving the trigonometry problems. In addition, I have composed the problems to mimic the College Board examples both in wording and difficulty.

After the seven chapters dealing with the various math topics, there are two randomized problem sets that include a mixture of all the problems. After going through the math topics separately, I recommend the reader try one of the problem sets to find out if there are still any of the types of problems that require more attention and drill.

Although I have endeavored to eradicate all errors in the manuscript, it would be foolish for me to bet I have actually created perfection. Therefore, I ask you, the reader and user of this book, to inform me of any errors you find and/or to request a current errata sheet. Please e-mail me at richard.shedenhelm@gmail.com.

Athens, Georgia, May 15, 2017.

Richard Shedenhelm

1. Solving Simple Equations

The College Board presents problems involving solving simple equations in three formats. Here is one example of each format:

1. If $\frac{x-1}{5} = k$ and $k = 5$, what is the value of x ?

2.

$$\sqrt{x-a} = x - 5$$

If $a = 3$, what is the solution set of the equation above?

3. If $a = 6\sqrt{2}$ and $2a = \sqrt{2x}$, what is the value of x ?

There are two steps to solve the first problem: 1. Substitute the given value of k into the equation; 2. Solve for x. To wit:

$$\frac{x-1}{5} = 5$$

$$x - 1 = 25 \quad \Longrightarrow \quad x = 26$$

The second problem has three steps: 1. Substitute the given value of a into the equation; 2. Solve for x; 3. Test the solutions in the original equations to check for extraneous solutions. Following this procedure, we have:

$$\sqrt{x-3} = x - 5$$

$$\left(\sqrt{x-3}\right)^2 = (x-5)^2 \quad \Longrightarrow \quad x - 3 = x^2 - 10x + 25 \quad \Longrightarrow \quad 0 = x^2 - 11x + 28$$

$$\Longrightarrow \quad 0 = (x-7)(x-4) \quad \Longrightarrow \quad x = 7 \text{ and } x = 4$$

Testing the solutions, we have

$$\sqrt{7-3} \overset{?}{\cong} 7 - 5$$

$$\sqrt{4} \overset{\checkmark}{\cong} 2$$

However,

$$\sqrt{4-3} \overset{?}{=} 4-5$$

$$\sqrt{1} \neq -1$$

Hence, the solution set must omit 4.

The third question has two steps: 1. Substitute the given value of a into the second equation; 2. Solve for x. This guidance gives us:

$$2\left(6\sqrt{2}\right) = \sqrt{2x}$$

$$12\sqrt{2} = \sqrt{2}\sqrt{x} \implies 12 = \sqrt{x} \implies 12^2 = \left(\sqrt{x}\right)^2 \implies 144 = x$$

PROBLEMS

1. If $\frac{x-1}{3} = k$ and $k = 3$, what is the value of x ?

2. If $\frac{x-1}{4} = k$ and $k = 5$, what is the value of x ?

3. If $\frac{x-1}{2} = k$ and $k = 4$, what is the value of x ?

4. If $\frac{x-1}{4} = k$ and $k = 4$, what is the value of x ?

5.
$$\sqrt{x-a} = x - 4$$

If $a = 2$, what is the solution set of the equation above?

6.
$$\sqrt{x-a} = x - 4$$

If $a = 4$, what is the solution set of the equation above?

7.
$$\sqrt{x-a} = x - 5$$

If $a = -15$, what is the solution set of the equation above?

8.
$$\sqrt{x-a} = x - 3$$

If $a = 1$, what is the solution set of the equation above?

9. If $a = 5\sqrt{2}$ and $3a = \sqrt{2x}$, what is the value of x ?

10. If $a = 7\sqrt{2}$ and $3a = \sqrt{2x}$, what is the value of x ?

11. If $a = 3\sqrt{3}$ and $3a = \sqrt{3x}$, what is the value of x ?

12. If $a = 4\sqrt{2}$ and $2a = \sqrt{2x}$, what is the value of x ?

ANSWERS

1. 10	5. {6}	9. 225
2. 21	6. {4, 5}	10. 441
3. 9	7. {10}	11. 81
4. 17	8. {5}	12. 64

SELECTED SOLUTIONS

1.   There are two steps to solve the first problem: 1. Substitute the given value of k into the equation; 2. Solve for x. To wit:

$$\frac{x - 1}{3} = 3$$

$$x - 1 = 9 \quad \Rightarrow \quad x = 10$$

5.   The second problem has three steps: 1. Substitute the given value of a into the equation; 2. Solve for x; 3. Test the solutions in the original equations to check for extraneous solutions. Following this procedure, we have:

$$\sqrt{x - 2} = x - 4$$

$$\left(\sqrt{x - 2}\right)^2 = (x - 4)^2 \quad \Rightarrow \quad x - 2 = x^2 - 8x + 16 \quad \Rightarrow \quad 0 = x^2 - 9x + 18$$

$$\Rightarrow \quad 0 = (x - 6)(x - 3) \quad \Rightarrow \quad x = 6 \text{ and } x = 3$$

Testing the solutions, we have

$$\sqrt{6 - 2} \overset{?}{=} 6 - 4$$

$$\sqrt{4} \overset{\checkmark}{=} 2$$

However,

$$\sqrt{3 - 2} \overset{?}{=} 3 - 4$$

5

$$\sqrt{1} \neq -1$$

Hence, the solution set must omit 3.

9.   The third question has two steps: 1. Substitute the given value of a into the second equation; 2. Solve for x. This guidance gives us:

$$3\left(5\sqrt{2}\right) = \sqrt{2x}$$

$$15\sqrt{2} = \sqrt{2}\sqrt{x} \quad \Longrightarrow \quad 15 = \sqrt{x} \quad \Longrightarrow \quad 15^2 = \left(\sqrt{x}\right)^2 \quad \Longrightarrow \quad 225 = x$$

2. Solving Quadratic Equations

The College Board presents problems involving quadratic equations in seven formats. Here is one example of each format:

1. What are the solutions to $2x^2 + 8x + 4 = 0$?

2. If $x > 0$ and $x^2 + 3x - 4 = 0$, what is the value of x?

3. What is the sum of all values of m that satisfy $3m^2 - 12m + 6 = 0$?

4.
$$2x^2 + 13x - 15 = 0$$

If r and s are two solutions of the equation above and $r > s$, what is the value of $r - s$?

5.
$$x^2 - \frac{k}{3}x = 3p$$

In the quadratic equation above, k and p are constants. What are the solutions for x ?

A) $x = \frac{k}{6} \pm \frac{\sqrt{k^2 + 2p}}{6}$

B) $x = \frac{k}{6} \pm \frac{\sqrt{k^2 + 108p}}{6}$

C) $x = \frac{k}{3} \pm \frac{\sqrt{k^2 + 2p}}{6}$

D) $x = \frac{k}{3} \pm \frac{\sqrt{k^2 + 108p}}{6}$

6.

$$(x + 3)^2 - 16 = 0$$

What is a value of x that satisfies the equation above?

7. What are the solutions to the equation

$$2x^2 - 50 = 0 \ ?$$

The first observation to make about these problems is that the quadratic formula either can or must be used for each of the problems. As a reminder, for a quadratic equation of the form

$$ax^2 + bx + c = 0 \quad (a \neq 0)$$

the solutions (also known as "the zeros" or "the roots") are found by the quadratic formula

$$x = \frac{-b \pm \sqrt{b^2 - 4ac}}{2a}.$$

The last two problems, however, can be solved by different methods.

Now, down to the details:

1. Using the quadratic formula, with $a = 2$, $b = 8$, and $c = 4$, we have

$$x = \frac{-8 \pm \sqrt{8^2 - 4(2)(4)}}{2(2)} = \frac{-8 \pm \sqrt{64 - 32}}{4} = \frac{-8 \pm \sqrt{32}}{4} = \frac{-8 \pm \sqrt{16 \cdot 2}}{4} = \frac{-8 \pm \sqrt{4^2 \cdot 2}}{4} =$$

$$= \frac{-8 \pm 4\sqrt{2}}{4} = \frac{4\left(-2 \pm \sqrt{2}\right)}{4} = -2 \pm \sqrt{2}.$$

Therefore, the solutions are $x = -2 + \sqrt{2}$ and $x = -2 - \sqrt{2}$.

The biggest source of error in problems of this sort have to do with cancellation. For example, if we had solutions looking like $x = \frac{-8 \pm 5\sqrt{2}}{4}$, we could *not* simplify the fraction to $-2 \pm 5\sqrt{2}$, since there is not a common factor of 4 in *both the terms of the numerator*. Also note that the College Board always reduces radicals to lowest terms, e.g., $2\sqrt{2}$, not $\sqrt{8}$.

2. Using the quadratic formula, with $a = 1$, $b = 3$, and $c = -4$, we have

$$x = \frac{-3 \pm \sqrt{(-3)^2 - 4(1)(-4)}}{2(1)} = \frac{-3 \pm \sqrt{9 + 16}}{2} = \frac{-3 \pm \sqrt{25}}{2} = \frac{-3 \pm \sqrt{5^2}}{2} = \frac{-3 \pm 5}{2}.$$

Therefore, the solutions are $x = \frac{-3+5}{2} = \frac{2}{2} = 1$ and $= \frac{-3-5}{2} = \frac{-8}{2} = -4$. However, with the constraint in the original problem that $x > 0$, the only final solution is $x = 1$.

3. Using the quadratic formula, with $a = 3$, $b = -12$, and $c = 6$, we have

$$m = \frac{-(-12) \pm \sqrt{(-12)^2 - 4(3)(6)}}{2(3)} = \frac{12 \pm \sqrt{144 - 72}}{6} = \frac{12 \pm \sqrt{72}}{6} = \frac{12 \pm \sqrt{36 \cdot 2}}{6} =$$

$$= \frac{12 \pm \sqrt{6^2 \cdot 2}}{6} = \frac{12 \pm 6\sqrt{2}}{6} = \frac{6\left(2 \pm \sqrt{2}\right)}{6} = 2 \pm \sqrt{2}.$$

Therefore, the solutions are $m = 2 + \sqrt{2}$ and $= 2 - \sqrt{2}$. However, the problem asks for the *sum* of the two solutions. Hence, the final answer is $\left(2 + \sqrt{2}\right) + \left(2 - \sqrt{2}\right) =$ $= 2 + \sqrt{2} + 2 - \sqrt{2} = 4$.

4. Using the quadratic formula, with $a = 2$, $b = 13$, and $c = -15$, we have

$$x = \frac{-13 \pm \sqrt{(-13)^2 - 4(2)(-15)}}{2(2)} = \frac{-13 \pm \sqrt{169 + 120}}{4} = \frac{-13 \pm \sqrt{289}}{4} = \frac{-13 \pm \sqrt{17^2}}{4} =$$

$$= \frac{-13 \pm 17}{4}.$$

Therefore, the solutions are $x = \frac{-13+17}{4} = \frac{4}{4} = 1$ and $= \frac{-13-17}{4} = \frac{-30}{4} = \frac{-15}{2}$. Since $1 > \frac{-15}{2}$, $r = 1$ and $s = \frac{-15}{2}$. The problem asks for the positive *difference* of the two solutions, viz., $r - s$. Hence, the final answer is $1 - \left(\frac{-15}{2}\right) = 1 + \frac{15}{2} = \frac{2}{2} + \frac{15}{2} = \frac{17}{2}$.

5. Before using the quadratic formula, we need to arrange the terms of the equation in the standard order, in order to correctly identify the constants a, b, and c. Hence, we have

$$x^2 - \frac{k}{3}x - 3p = 0 .$$

We will make the problem easier to solve if we eliminate the fraction in the linear term. We can accomplish this by multiplying each term by 3. So, we will have

$$3x^2 - kx - 9p = 0 .$$

Now, using the quadratic formula, with $a = 3$, $b = -k$, and $c = -9p$, we have

$$x = \frac{-(-k) \pm \sqrt{(-k)^2 - 4(3)(-9p)}}{2(3)} = \frac{k \pm \sqrt{k^2 + 108p}}{6} = \frac{k}{6} \pm \frac{\sqrt{k^2 + 108p}}{6}.$$

Therefore, the answer is option B.

6. This problem can be solved using the quadratic formula, if we first expand the first term and then collect like terms, viz.,

$$(x + 3)^2 - 16 = 0 \implies x^2 + 6x + 9 - 16 = 0 \implies x^2 + 6x - 7 = 0.$$

Using the quadratic formula, with $a = 1$, $b = 6$, and $c = -7$, we have

$$x = \frac{-6 \pm \sqrt{6^2 - 4(1)(-7)}}{2(1)} = \frac{-6 \pm \sqrt{36 + 28}}{2} = \frac{-6 \pm \sqrt{64}}{2} = \frac{-6 \pm 8}{2} = \frac{2(-3 \pm 4)}{2} =$$

$$= -3 \pm 4.$$

Therefore, the solutions are $x = -3 + 4 = 1$ and $x = -3 - 4 = -7$.

An alternative way to solve this problem—not using the quadratic formula—goes as follows.

$$(x + 3)^2 - 16 = 0 \implies (x + 3)^2 = 16 \implies \sqrt{(x + 3)^2} = \pm\sqrt{16} \implies x + 3 = \pm 4$$

$$\implies x = -3 \pm 4.$$

Therefore, as before, the solutions are $x = -3 + 4 = 1$ and $x = -3 - 4 = -7$.

7. This problem can be solved using the quadratic formula, just by letting $b = 0$. Hence, using the quadratic formula, with $a = 2$, $b = 0$, and $c = -50$, we have

$$x = \frac{-0 \pm \sqrt{0^2 - 4(2)(-50)}}{2(2)} = \frac{\pm\sqrt{400}}{4} = \frac{\pm 20}{4} = \pm 5.$$

Therefore, the solutions are $x = 5$ and $x = -5$.

An alternative way to solve this problem—not using the quadratic formula—goes as follows.

$$2x^2 - 50 = 0 \implies 2x^2 = 50 \implies x^2 = 25 \implies \sqrt{x^2} = \pm 5 \implies x = \pm 5.$$

The advantage to doing problems like 6 and 7 using the quadratic formula is that we will have a single method to solve *all* questions requiring us to solve quadratic equations. In addition, many students make the mistake of forgetting the "$\pm$" in the square-rooting step of the alternative method, e.g., $(x + 3)^2 = 16 \implies \sqrt{(x + 3)^2} = \pm\sqrt{16}$.

PROBLEMS

1. What are the solutions to $2x^2 + 8x + 2 = 0$?

2. What are the solutions to $3x^2 + 12x + 6 = 0$?

3. What are the solutions to $4x^2 + 16x + 8 = 0$?

4. If $x > 0$ and $5x^2 + 4x - 1 = 0$, what is the value of x?

5. If $x > 0$ and $3x^2 + 5x - 2 = 0$, what is the value of x?

6. If $x > 0$ and $3x^2 - 4x + 1 = 0$, what is the value of x?

7. What is the sum of all values of m that satisfy $3m^2 - 12m + 3 = 0$?

8. What is the sum of all values of m that satisfy $m^2 - 8m + 4 = 0$?

9. What is the sum of all values of m that satisfy $2m^2 - 18m + 3 = 0$?

10.
$$3x^2 + 6x - 9 = 0$$

If r and s are two solutions of the equation above and $r > s$, what is the value of $r - s$?

11.

$$5x^2 + 7x - 6 = 0$$

If r and s are two solutions of the equation above and $r > s$, what is the value of $r - s$?

12.

$$3x^2 + 8x - 11 = 0$$

If r and s are two solutions of the equation above and $r > s$, what is the value of $r - s$?

13.

$$x^2 - \frac{k}{2}x = 2p$$

In the quadratic equation above, k and p are constants. What are the solutions for x ?

A) $x = \frac{k}{4} \pm \frac{\sqrt{k^2+2p}}{4}$

B) $x = \frac{k}{2} \pm \frac{\sqrt{k^2+32p}}{4}$

C) $x = \frac{k}{4} \pm \frac{\sqrt{k^2+2p}}{8}$

D) $x = \frac{k}{4} \pm \frac{\sqrt{k^2+32p}}{4}$

14.

$$x^2 - \frac{k}{4}x = 4p$$

In the quadratic equation above, k and p are constants. What are the solutions for x ?

A) $x = \frac{k}{4} \pm \frac{\sqrt{k^2+4p}}{4}$

B) $x = \frac{k}{2} \pm \frac{\sqrt{k^2+4p}}{4}$

C) $x = \frac{k}{8} \pm \frac{\sqrt{k^2+256p}}{8}$

D) $x = \frac{k}{4} \pm \frac{\sqrt{k^2+256p}}{4}$

15.

$$x^2 - \frac{k}{2}x = 3p$$

In the quadratic equation above, k and p are constants. What are the solutions for x ?

A) $x = \frac{k}{2} \pm \frac{\sqrt{k^2+48p}}{2}$

B) $x = \frac{k}{2} \pm \frac{\sqrt{k^2+3p}}{2}$

C) $x = \frac{k}{3} \pm \frac{\sqrt{k^2+2p}}{3}$

D) $x = \frac{k}{4} \pm \frac{\sqrt{k^2+48p}}{4}$

16.

$$(x+4)^2 - 9 = 0$$

What is a value of x that satisfies the equation above?

17.

$$(x+2)^2 - 9 = 0$$

What is a value of x that satisfies the equation above?

18.

$$(x+2)^2 - 25 = 0$$

What is a value of x that satisfies the equation above?

19. What are the solutions to the equation
$$2x^2 - 72 = 0 \text{ ?}$$

20. What are the solutions to the equation
$$2x^2 - 32 = 0 \text{ ?}$$

21. What are the solutions to the equation
$$3x^2 - 75 = 0 \text{ ?}$$

ANSWERS

1. $-2 \pm \sqrt{3}$	8. 8	15. D
2. $-2 \pm \sqrt{2}$	9. 9	16. -7 and -1
3. $-2 \pm \sqrt{2}$	10. 4	17. -5 and 1
4. $\frac{1}{5}$	11. $\frac{13}{5}$	18. -7 and 3
5. $\frac{1}{3}$	12. $\frac{14}{3}$	19. -6 and 6
6. $\frac{1}{3}$ or 1	13. D	20. -4 and 4
7. 4	14. C	21. -5 and 5

SELECTED SOLUTIONS

1. Using the quadratic formula, with $a = 2$, $b = 8$, and $c = 2$, we have

$$x = \frac{-8 \pm \sqrt{8^2 - 4(2)(2)}}{2(2)} = \frac{-8 \pm \sqrt{64 - 16}}{4} = \frac{-8 \pm \sqrt{48}}{4} = \frac{-8 \pm \sqrt{16 \cdot 3}}{4} = \frac{-8 \pm \sqrt{4^2 \cdot 3}}{4} =$$

$$= \frac{-8 \pm 4\sqrt{3}}{4} = \frac{4\left(-2 \pm \sqrt{3}\right)}{4} = -2 \pm \sqrt{3}.$$

Therefore, the solutions are $x = -2 + \sqrt{3}$ and $x = -2 - \sqrt{3}$.

Alternative solution:

$$2x^2 + 8x + 2 = 0 \implies 2(x^2 + 4x + 1) = 0 \implies x^2 + 4x + 1 = 0.$$

Using the quadratic formula, with $a = 1$, $b = 4$, and $c = 1$, we have

$$x = \frac{-4 \pm \sqrt{4^2 - 4(1)(1)}}{2(1)} = \frac{-4 \pm \sqrt{16 - 4}}{2} = \frac{-4 \pm \sqrt{12}}{2} = \frac{-4 \pm \sqrt{4 \cdot 3}}{2} = \frac{-4 \pm \sqrt{2^2 \cdot 3}}{2} =$$

$$= \frac{-4 \pm 2\sqrt{3}}{2} = \frac{2\left(-2 \pm \sqrt{3}\right)}{2} = -2 \pm \sqrt{3}.$$

Therefore, as before, the solutions are $x = -2 + \sqrt{3}$ and $x = -2 - \sqrt{3}$.

4. Using the quadratic formula, with $a = 5$, $b = 4$, and $c = -1$, we have

$$x = \frac{-4 \pm \sqrt{(4)^2 - 4(5)(-1)}}{2(5)} = \frac{-4 \pm \sqrt{16 + 20}}{10} = \frac{-4 \pm \sqrt{36}}{10} = \frac{-4 \pm \sqrt{6^2}}{10} = \frac{-4 \pm 6}{10}.$$

Therefore, the solutions are $x = \frac{-4+6}{10} = \frac{2}{10} = \frac{1}{5}$ and $= \frac{-4-6}{10} = \frac{-10}{10} = -1$. However, with the constraint in the original problem that $x > 0$, the only final solution is $x = \frac{1}{5}$.

7. Using the quadratic formula, with $a = 3$, $b = -12$, and $c = 3$, we have

$$m = \frac{-(-12) \pm \sqrt{(-12)^2 - 4(3)(3)}}{2(3)} = \frac{12 \pm \sqrt{144 - 36}}{6} = \frac{12 \pm \sqrt{108}}{6} = \frac{12 \pm \sqrt{36 \cdot 3}}{6} =$$

$$= \frac{12 \pm \sqrt{6^2 \cdot 3}}{6} = \frac{12 \pm 6\sqrt{3}}{6} = \frac{6(2 \pm \sqrt{3})}{6} = 2 \pm \sqrt{3}.$$

Therefore, the solutions are $m = 2 + \sqrt{3}$ and $= 2 - \sqrt{3}$. However, the problem asks for the *sum* of the two solutions. Hence, the final answer is $(2 + \sqrt{3}) + (2 - \sqrt{3}) =$
$= 2 + \sqrt{3} + 2 - \sqrt{3} = 4$.

Alternative solution:

$$3m^2 - 12m + 3 = 0 \quad \Rightarrow \quad 3(m^2 - 4m + 1) = 0 \quad \Rightarrow \quad m^2 - 4m + 1 = 0.$$

Using the quadratic formula, with $a = 1$, $b = -4$, and $c = 1$, we have

$$x = \frac{-(-4) \pm \sqrt{(-4)^2 - 4(1)(1)}}{2(1)} = \frac{4 \pm \sqrt{16 - 4}}{2} = \frac{4 \pm \sqrt{12}}{2} = \frac{4 \pm \sqrt{4 \cdot 3}}{2} = \frac{4 \pm \sqrt{2^2 \cdot 3}}{2} =$$

$$= \frac{4 \pm 2\sqrt{3}}{2} = \frac{2(2 \pm \sqrt{3})}{2} = 2 \pm \sqrt{3}.$$

Therefore, as before, the initial solutions are $x = 2 + \sqrt{3}$ and $x = 2 - \sqrt{3}$.

10. Using the quadratic formula, with $a = 3$, $b = 6$, and $c = -9$, we have

$$x = \frac{-6 \pm \sqrt{(6)^2 - 4(3)(-9)}}{2(3)} = \frac{-6 \pm \sqrt{36 + 108}}{6} = \frac{-6 \pm \sqrt{144}}{6} = \frac{-6 \pm \sqrt{12^2}}{6} =$$

$$= \frac{-6 \pm 12}{6}.$$

Therefore, the solutions are $x = \frac{-6+12}{6} = \frac{6}{6} = 1$ and $= \frac{-6-12}{6} = \frac{-18}{6} = -3$. However, the problem asks for the positive *difference* of the two solutions. Hence, the final answer is

$1 - (-3) = 1 + 3 = 4$.

Alternative solution:

$3x^2 + 6x - 9 = 0 \implies 3(x^2 + 2x - 3) = 0 \implies x^2 + 2x - 3 = 0.$

Using the quadratic formula, with $a = 1$, $b = 2$, and $c = -3$, we have

$$x = \frac{-2 \pm \sqrt{2^2 - 4(1)(-3)}}{2(1)} = \frac{-2 \pm \sqrt{4 + 12}}{2} = \frac{-2 \pm \sqrt{16}}{2} = \frac{-2 \pm \sqrt{4^2}}{2} =$$

$$= \frac{-2 \pm 4}{2} = \frac{2(-1 \pm 2)}{2} = -1 \pm 2.$$

Therefore, as before, the initial solutions are $x = -1 + 2 = 1$ and $x = -1 - 2 = -3$.

13.   Before using the quadratic formula, we need to arrange the terms of the equation in the standard order, in order to correctly identify the constants a, b, and c. Hence, we have

$$x^2 - \frac{k}{2}x - 2p = 0.$$

We will make the problem easier to solve if we eliminate the fraction in the linear term. We can accomplish this by multiplying each term by 2. So, we will have

$$2x^2 - kx - 4p = 0.$$

Now, using the quadratic formula, with $a = 2$, $b = -k$, and $c = -4p$, we have

$$x = \frac{-(-k) \pm \sqrt{(-k)^2 - 4(2)(-4p)}}{2(2)} = \frac{k \pm \sqrt{k^2 + 32p}}{4} = \frac{k}{4} \pm \frac{\sqrt{k^2 + 32p}}{4}.$$

Therefore, the answer is option D.

14.   Before using the quadratic formula, we need to arrange the terms of the equation in the standard order, in order to correctly identify the constants a, b, and c. Hence, we have

$$x^2 - \frac{k}{4}x - 4p = 0.$$

We will make the problem easier to solve if we eliminate the fraction in the linear term. We can accomplish this by multiplying each term by 4. So, we will have

$$4x^2 - kx - 16p = 0.$$

Now, using the quadratic formula, with $a = 4$, $b = -k$, and $c = -16p$, we have

$$x = \frac{-(-k) \pm \sqrt{(-k)^2 - 4(4)(-16p)}}{2(4)} = \frac{k \pm \sqrt{k^2 + 256p}}{8} = \frac{k}{8} \pm \frac{\sqrt{k^2 + 256p}}{8}.$$

Therefore, the answer is option C.

15.   Before using the quadratic formula, we need to arrange the terms of the equation in the standard order, in order to correctly identify the constants a, b, and c. Hence, we have

$$x^2 - \frac{k}{2}x - 3p = 0.$$

We will make the problem easier to solve if we eliminate the fraction in the linear term. We can accomplish this by multiplying each term by 2. So, we will have

$$2x^2 - kx - 6p = 0.$$

Now, using the quadratic formula, with $a = 2$, $b = -k$, and $c = -6p$, we have

$$x = \frac{-(-k) \pm \sqrt{(-k)^2 - 4(2)(-6p)}}{2(2)} = \frac{k \pm \sqrt{k^2 + 48p}}{4} = \frac{k}{4} \pm \frac{\sqrt{k^2 + 48p}}{4}.$$

Therefore, the answer is option D.

16.   This problem can be solved using the quadratic formula, if we first expand the first term and then collect like terms, viz.,

$$(x + 4)^2 - 9 = 0 \quad \Longrightarrow \quad x^2 + 8x + 16 - 9 = 0 \quad \Longrightarrow \quad x^2 + 8x + 7 = 0.$$

Using the quadratic formula, with $a = 1$, $b = 8$, and $c = 7$, we have

$$x = \frac{-8 \pm \sqrt{8^2 - 4(1)(7)}}{2(1)} = \frac{-8 \pm \sqrt{64 - 28}}{2} = \frac{-8 \pm \sqrt{36}}{2} = \frac{-8 \pm 6}{2} = \frac{2(-4 \pm 3)}{2} =$$

$$= -4 \pm 3.$$

Therefore, the solutions are $x = -4 + 3 = -1$ and $x = -4 - 3 = -7$.

Alternative solution:

$$(x + 4)^2 - 9 = 0 \implies (x + 4)^2 = 9 \implies \sqrt{(x + 4)^2} = \pm\sqrt{9} \implies x + 4 = \pm 3$$

$$\implies x = -4 \pm 3 \,.$$

Therefore, as before, the solutions are $x = -4 + 3 = -1$ and $x = -4 - 3 = -7$.

19.   This problem can be solved using the quadratic formula, just by letting $b = 0$. Hence, using the quadratic formula, with $a = 2$, $b = 0$, and $c = -72$, we have

$$x = \frac{-0 \pm \sqrt{0^2 - 4(2)(-72)}}{2(2)} = \frac{\pm\sqrt{576}}{4} = \frac{\pm 24}{4} = \pm 6.$$

Therefore, the solutions are $x = 6$ and $x = -6$.
    We can make the above approach to solving the problem easier if we first factor out the greatest common factor:

$$2x^2 - 72 = 0 \implies 2(x^2 - 36) = 0 \implies x^2 - 36 = 0.$$

Now we can use the quadratic formula with $a = 1$, $b = 0$, and $c = -36$. Hence, we have

$$x = \frac{-0 \pm \sqrt{0^2 - 4(1)(-36)}}{2(1)} = \frac{\pm\sqrt{144}}{2} = \frac{\pm 12}{2} = \pm 6.$$

Alternative solution:

$$2x^2 - 72 = 0 \implies 2x^2 = 72 \implies x^2 = 36 \implies \sqrt{x^2} = \pm 6 \implies x = \pm 6.$$

Therefore, as before, the solutions are $x = 6$ and $x = -6$.

3. Function Notation

A simple way to understand what functions do is to imagine a "rule machine" that takes in inputs and produces outputs. Let x represent the inputs and y the outputs. Now this machine produces the outputs according to a rule specified by an equation. For example, one rule machine could be:

$$x \Rightarrow \boxed{y = 2x + 3} \Rightarrow y$$

In words, we could say that the rule machine takes an input, multiplies it by 2, and then adds 3 to the product. Some examples of inputs becoming outputs would be:

$$1 \Rightarrow \boxed{y = 2(1) + 3} \Rightarrow 5$$

$$2 \Rightarrow \boxed{y = 2(2) + 3} \Rightarrow 7$$

$$3 \Rightarrow \boxed{y = 2(3) + 3} \Rightarrow 9$$

Typically the inputs and outputs are communicated as *ordered pairs* of the form (x, y). Hence, for the examples about we have $(1, 5)$, $(2, 7)$, and $(3, 9)$.

The only restriction on the function rule machine is that each input produces a *unique* output. When graphing functions, this restriction is often described as the "vertical line test." For example, the equation $y = x^2$ is a function, since it passes the vertical line test. However, the equation $x = y^2$ is not a function, since it fails the same test.

A last issue regarding the basic concept of functions concerns the letters standing for the inputs and outputs. Let us adopt the shorthand $x \to y$ to stand in place of the rule machine diagrams we used above. With that shorthand, the student will often see $t \to P$, where t (for time) is the input and P is the output. Other popular choices are $\theta \to y$ (often used in trigonometric applications) and $t \to x$.

The ordered pair (x, y) can be written using function notation, which consists of an equation such as $f(x) = y$. Using this equation, we can identify the three parts to this notation. From left to right, the first part is the name of the function. It can be a single letter or a whole word. The second part is the pair of parentheses, which contain the input(s). Each input is called an "argument" of the function. Finally, the third part—on the right side of the equality sign—is the output corresponding to the given input. Each such output is called a "value" of the function.

$$f(x) = y$$

name argument value

The College Board presents problems involving function notation in three formats. Here is one example of each format:

1. If $f(x) = -3x + 5$, what is $f(-2x)$ equal to?

2. If $g(x) = 3x + 1$ and $f(x) = g(x) + 4$, what is $f(2)$?

3.

$$f(x) = \frac{3}{2}x + b$$

In the function above, b is a constant. If $f(6) = 8$, what is the value of $f(-2)$?

To solve the first problem, we simply substitute the expression "$-2x$" in place of "x" in the given equation $f(x) = -3x + 5$. Hence, we get $f(-2x) = -3(-2x) + 5$, which simplifies to $f(-2x) = 6x + 5$. So, the answer is "$6x + 5$."

To answer the second problem, first note that $g(2) = 3(2) + 1$. Hence, by substitution, $f(2) = g(2) + 4 = [3(2) + 1] + 4 = 7 + 4 = 11$, viz., the answer is 11.

The last problem requires two steps. First, solving for the value of b. Second, using the value of b to find $f(-2)$. Since $f(6) = 8$, $8 = \frac{3}{2}(6) + b$. Hence, $8 = 3(3) + b$. So, $8 = 9 + b$. Thus, $-1 = b$. Now we can reformulate the function as $f(x) = \frac{3}{2}x - 1$. Hence, $f(-2) = \frac{3}{2}(-2) - 1 = -3 - 1 = -4$. Therefore, the answer is -4.

PROBLEMS

1. If $f(x) = -3x + 6$, what is $f(-2x)$ equal to?

2. If $f(x) = -2x + 7$, what is $f(-4x)$ equal to?

3. If $f(x) = 2x + 4$, what is $f(3x)$ equal to?

4. If $f(x) = -x + 5$, what is $f(-2x)$ equal to?

5. If $g(x) = 3x + 2$ and $f(x) = g(x) + 5$, what is $f(2)$?

6. If $g(x) = x + 1$ and $f(x) = g(x) + 3$, what is $f(3)$?

7. If $g(x) = -x + 2$ and $f(x) = g(x) + 4$, what is $f(5)$?

8. If $g(x) = -2x - 3$ and $f(x) = g(x) - 4$, what is $f(-3)$?

9.

$$f(x) = \frac{3}{2}x + b$$

In the function above, b is a constant. If $f(4) = 6$, what is the value of $f(-2)$?

10.

$$f(x) = \frac{5}{2}x + b$$

In the function above, b is a constant. If $f(6) = 8$, what is the value of $f(-4)$?

11.

$$f(x) = \frac{3}{4}x + b$$

In the function above, b is a constant. If $f(8) = 12$, what is the value of $f(4)$?

12.

$$f(x) = \frac{6}{8}x + b$$

In the function above, b is a constant. If $f(16) = -12$, what is the value of $f(-24)$?

1. $6x + 6$	5. 13	9. -3
2. $8x + 7$	6. 7	10. -17
3. $6x + 4$	7. 1	11. 9
4. $2x + 5$	8. -1	12. -42

SELECTED SOLUTIONS

1.   To solve this problem, we simply substitute the expression "$-2x$" in place of "x" in the given equation $f(x) = -3x + 6$. Hence, we get $f(-2x) = -3(-2x) + 6$, which simplifies to $f(-2x) = 6x + 6$. So, the answer is "$6x + 6$."

5.   To answer this problem, first note that $g(2) = 3(2) + 2$. Hence, by substitution, $f(2) = g(2) + 5 = [3(2) + 2] + 5 = 8 + 5 = 13$, viz., the answer is 13.

9.   This problem requires two steps. First, solving for the value of b. Second, using the value of b to find $f(-2)$. Since $f(4) = 6$, $6 = \frac{3}{2}(4) + b$. Hence, $6 = 3(2) + b$. So, $6 = 6 + b$. Thus, $0 = b$. Now we can reformulate the function as $f(x) = \frac{3}{2}x + 0$. Hence, $f(-2) = \frac{3}{2}(-2) + 0 = -3 + 0 = -3$. Therefore, the answer is -3.

12.   This problem requires two steps. First, solving for the value of b. Second, using the value of b to find $f(-24)$. Since $f(16) = -12$, $-12 = \frac{6}{8}(16) + b$. Hence, $-12 = 6(2) + b$. So, $-12 = 12 + b$. Thus, $-24 = b$. Now we can reformulate the function as $f(x) = \frac{6}{8}x - 24$. Hence, $f(-24) = \frac{6}{8}(-24) - 24 = -18 - 24 = -42$. Therefore, the answer is -42.

4. Systems of Linear Equations

The College Board presents problems involving systems of linear equations in seven formats. Here is one example of each format:

1. Solve the following system:

$$x + y = 1$$
$$5x - y = 23$$

2.

$$2x - 3y = -14$$
$$3x - 2y = -6$$

If (x, y) is a solution to the system of equations above, what is the value of $x - y$?

A) -20

B) -8

C) -4

D) 8

3.

$$3x + b = 5x - 7$$
$$3y + c = 5y - 7$$

In the equations above, b and c are constants. If b is c minus $\frac{1}{2}$, which of the following is true?

A) x is y minus $\frac{1}{4}$.

B) x is y minus $\frac{1}{2}$.

C) x is y minus 1.

D) x is y plus $\frac{1}{2}$.

4.

$$ax + by = 12$$
$$2x + 8y = 60$$

In the system of equations above, a and b are constants. If the system has infinitely many solutions, what is the value of $\frac{a}{b}$?

5.

$$kx - 3y = 4$$
$$4x - 5y = 7$$

In the system of equations above, k is a constant and x and y are variables. For what value of k will the system of equations have no solution?

6. Which of the following equations represents a line that is parallel to the line with equation $y = -3x + 4$?

A) $6x + 2y = 15$
B) $3x - y = 7$
C) $2x - 3y = 6$
D) $x + 3y = 1$

7.

$$y = x - 4$$
$$4x - 4y = 12$$

The system of equations above consists of two equations, and the graph of each equation in the xy-plane is a line. Which of the following statements is true about these two lines?

A) The lines are parallel.

B) The lines are the same.

C) The lines are perpendicular.

D) The lines intersect at $(-4, -3)$.

Now for the solutions:
1. Using the "elimination method," we have

$$\begin{array}{r} x + y = 1 \\ \underline{5x - y = 23} \\ 6x \quad\; = 24 \end{array}$$

Hence,

$$\frac{6x}{6} = \frac{24}{6}.$$

So,

$$x = 4.$$

Thus,

$$4 + y = 1 \quad \Longrightarrow \quad y = 1 - 4 = -3.$$

Alternative solution:

Using the "substitution method," we could begin as follows:

$$y = 1 - x.$$

Hence,

$$5x - (1 - x) = 23,$$

by substitution. So,

$$5x - 1 + x = 23 \quad \Longrightarrow \quad 6x = 24 \quad \Longrightarrow \quad x = 4.$$

Thus,

$$y = 1 - 4 = -3.$$

2. Using the "elimination method," we have

$$\begin{array}{c} (-2)(2x - 3y) = -14(-2) \\ (3)(3x - 2y) = -6(3) \end{array}$$

Hence,

$$\begin{array}{r} -4x + 6y = 28 \\ \underline{9x - 6y = -18} \\ 5x \quad\quad = 10 \end{array}$$

So,

$$\frac{5x}{5} = \frac{10}{5}.$$

Thus,

$$x = 2.$$

Hence,

$$2(2) - 3y = -14 \quad \Longrightarrow \quad 4 - 3y = -14 \quad \Longrightarrow \quad -3y = -18.$$

So,

$$y = 6.$$

Therefore,

$$x - y = 2 - 6 = -4,$$

which is option C.

3. First note that the problem states $b = c - \frac{1}{2}$. Hence, by substitution into the first equation,
$$3x + \left(c - \frac{1}{2}\right) = 5x - 7.$$
So,
$$c = 2x - 7 + \frac{1}{2} = 2x - \frac{14}{2} + \frac{1}{2} = 2x - \frac{13}{2}.$$
Thus, by substitution into the second equation,
$$3y + \left(2x - \frac{13}{2}\right) = 5y - 7.$$
Hence,
$$2x = 2y - 7 + \frac{13}{2} = 2y - \frac{14}{2} + \frac{13}{2} = 2y - \frac{1}{2}.$$
So,
$$x = y - \frac{1}{4},$$
which is option A.

4. The condition that the system has infinitely many solutions implies that the two equations are really the same! Multiplying the first equation by 5, the system becomes
$$5ax + 5by = 60$$
$$2x + 8y = 60$$

Hence,
$$5a = 2 \text{ and } 5b = 8.$$
So,
$$a = \frac{2}{5} \text{ and } b = \frac{8}{5}.$$
Therefore,
$$\frac{a}{b} = \frac{\left(\frac{2}{5}\right)}{\left(\frac{8}{5}\right)} = \frac{2}{5} \cdot \frac{5}{8} = \frac{2}{8} = \frac{1}{4}.$$

5. The condition that the system has no solution implies that the two equations represent two distinct but parallel lines. Hence, the slopes of the lines are the same. Converting the equations into slope-intercept form, we have

$$kx - 3y = 4 \implies kx - 4 = 3y \implies \frac{k}{3}x - \frac{4}{3} = y \implies y = \frac{k}{3}x - \frac{4}{3}$$

and

$$4x - 5y = 7 \implies 4x - 7 = 5y \implies \frac{4}{5}x - \frac{7}{5} = y \implies y = \frac{4}{5}x - \frac{7}{5}.$$

(Note that the equations have distinct y-intercepts.) So,

$$\frac{k}{3} = \frac{4}{5}.$$

Thus,

$$k = \frac{12}{5}$$

is the value of k that will render the system of equations to have no solution.

6. Since we want the equation that represents a line parallel to the equation

$$y = -3x + 4,$$

we need to find the equation that represents a line with the same slope, viz., $m = -3$. Converting equation A into slope-intercept form, we have

$$6x + 2y = 15 \implies 2y = -6x + 15 \implies y = -3x + \frac{15}{2}.$$

Hence, equation A has the slope $m = -3$. So, option A is the correct answer.

7. Converting the second equation into slope-intercept form, we have

$$4x - 4y = 12 \implies 4x - 12 = 4y \implies x - 3 = y \implies y = x - 3.$$

Hence, the slope of both lines is $m = 1$. However, the y-intercepts of the two lines are not the same. Therefore, we have two distinct but parallel lines, viz., option A.

31

PROBLEMS

Solve the following systems:

1. $x + y = 2$
    $5x - y = 22$

2. $-3x + 4y = 20$
    $6x + 3y = 15$

3. $2x - y = 6$
    $x + 2y = -2$

4. $x + y = -9$
    $x + 2y = -25$

5. $x + y = 0$
    $3x - 2y = 10$

6. $3x + 2y = 9$
    $5x - y = -11$

7. $\dfrac{1}{2}x - \dfrac{1}{4}y = 10$
    $\dfrac{1}{8}x - \dfrac{1}{8}y = 19$

8. $\dfrac{x}{y} = 6$
    $4(y + 1) = x$

9. $3x + 4y = -23$
    $2y - x = -19$

10. $-2x = 4y + 6$
    $2(2y + 3) = 3x - 5$

11.

$3x - 4y = -11$
$4x - 3y = 4$

If (x, y) is a solution to the system of equations above, what is the value of $x - y$?

A) -15

B) -7

C) -1

D) 7

12.

$$2x + 3y = 16$$
$$3x - 2y = -2$$

If (x, y) is a solution to the system of equations above, what is the value of $x - y$?

A) 14

B) −18

C) 0

D) −2

13.

$$2x + b = 4x - 6$$
$$2y + c = 4y - 6$$

In the equations above, b and c are constants. If b is c minus $\frac{1}{2}$, which of the following is true?

A) x is y plus $\frac{1}{4}$.

B) x is y minus $\frac{1}{2}$.

C) x is y minus 1.

D) x is y minus $\frac{1}{4}$.

14.

$$3x + b = 5x - 7$$
$$3y + c = 5y - 7$$

In the equations above, b and c are constants. If b is c minus $\frac{1}{4}$, which of the following is true?

A) x is y minus $\frac{1}{4}$.

B) x is y plus $\frac{1}{2}$.

C) x is y minus $\frac{1}{8}$.

D) x is y minus 1.

15.

$$ax + by = 9$$
$$3x + 4y = 54$$

In the system of equations above, a and b are constants. If the system has infinitely many solutions, what is the value of $\frac{a}{b}$?

16.

$$ax + by = 11$$
$$2x + 6y = 77$$

In the system of equations above, a and b are constants. If the system has infinitely many solutions, what is the value of $\frac{a}{b}$?

17.

$$kx - 2y = 5$$
$$3x - 4y = 8$$

In the system of equations above, k is a constant and x and y are variables. For what value of k will the system of equations have no solution?

18.

$$kx - 5y = 3$$
$$6x - 7y = 6$$

In the system of equations above, k is a constant and x and y are variables. For what value of k will the system of equations have no solution?

19. Which of the following equations represents a line that is parallel to the line with equation $y = -4x + 4$?

A) $6x + 4y = 15$
B) $4x - y = 7$
C) $8x + 2y = 6$
D) $x + 2y = 1$

20. Which of the following equations represents a line that is parallel to the line with equation $y = 2x + 3$?

A) $6x + 4y = 3$
B) $8x - 4y = 7$
C) $8x + 2y = 7$
D) $x + 6y = 10$

21.

$$y = x - 3$$
$$2y - 2x = 6$$

The system of equations above consists of two equations, and the graph of each equation in the xy-plane is a line. Which of the following statements is true about these two lines?

A) The lines are parallel.

B) The lines are the same.

C) The lines are perpendicular.

D) The lines intersect at $(-3, 6)$.

22.

$$y = x + 5$$
$$3x - 4y = 10$$

The system of equations above consists of two equations, and the graph of each equation in the xy-plane is a line. Which of the following statements is true about these two lines?

A) The lines are parallel.

B) The lines are the same.

C) The lines are perpendicular.

D) The lines intersect at $(-30, -25)$.

23.

$$y = 2x - 6$$
$$x + 2y = 16$$

The system of equations above consists of two equations, and the graph of each equation in the xy-plane is a line. Which of the following statements is true about these two lines?

A) The lines are parallel.

B) The lines are the same.

C) The lines are perpendicular.

D) The lines intersect at $(2, 2)$.

24.

$$y - x = 2$$
$$3x - 3y = 9$$

The system of equations above consists of two equations, and the graph of each equation in the xy-plane is a line. Which of the following statements is true about these two lines?

A) The lines are parallel.

B) The lines are the same.

C) The lines are perpendicular.

D) The lines intersect at $(2, 9)$.

25.

$$y = x - 5$$
$$-5x - 5y = 10$$

The system of equations above consists of two equations, and the graph of each equation in the xy-plane is a line. Which of the following statements is true about these two lines?

A) The lines are parallel.

B) The lines are the same.

C) The lines are perpendicular.

D) The lines intersect at $(-5, 10)$.

ANSWERS

1. $(4,-2)$	6. $(-1,6)$	11. C	16. $\frac{1}{3}$	21. A
2. $(0,5)$	7. $(-112,-264)$	12. D	17. $\frac{3}{2}$	22. D
3. $(2,-2)$	8. $(12,2)$	13. D	18. $\frac{30}{7}$	23. C
4. $(7,-16)$	9. $(3,-8)$	14. C	19. C	24. A
5. $(2,-2)$	10. $(1,-2)$	15. $\frac{3}{4}$	20. B	25. C

SELECTED SOLUTIONS

1.   Using the "elimination method," we have

$$\begin{aligned} x + y &= 2 \\ \underline{5x - y} &= \underline{22} \\ 6x &= 24 \end{aligned}$$

Hence,

$$\frac{6x}{6} = \frac{24}{6}.$$

So,

$$x = 4.$$

Thus,

$$4 + y = 2 \quad \Longrightarrow \quad y = 2 - 4 = -2.$$

Alternative solution:

Using the "substitution method," we could begin as follows:

$$y = 2 - x.$$

Hence,

$$5x - (2 - x) = 22,$$

by substitution. So,

$$5x - 2 + x = 22 \quad \Longrightarrow \quad 6x = 24 \quad \Longrightarrow \quad x = 4.$$

Thus,

$$y = 2 - 4 = -2.$$

38

11.   Using the "elimination method," we have

$$(-3)(3x - 4y) = -11(-3)$$
$$(4)(4x - 3y) = 4(4)$$

Hence,

$$-9x + 12y = 33$$
$$\underline{16x - 12y = 16}$$
$$7x \quad\quad = 49$$

So,

$$\frac{7x}{7} = \frac{49}{7}.$$

Thus,

$$x = 7.$$

Hence,

$$3(7) - 4y = -11 \quad\Rightarrow\quad 21 - 4y = -11 \quad\Rightarrow\quad -4y = -32.$$

So,

$$y = 8.$$

Therefore,

$$x - y = 7 - 8 = -1,$$

which is option C.

13.   First note that the problem states $b = c - \frac{1}{2}$. Hence, by substitution into the first equation,

$$2x + \left(c - \frac{1}{2}\right) = 4x - 6.$$

So,

$$c = 2x - 6 + \frac{1}{2} = 2x - \frac{12}{2} + \frac{1}{2} = 2x - \frac{11}{2}.$$

Thus, by substitution into the second equation,

$$2y + \left(2x - \frac{11}{2}\right) = 4y - 6.$$

Hence,

$$2x = 2y - 6 + \frac{11}{2} = 2y - \frac{12}{2} + \frac{11}{2} = 2y - \frac{1}{2}.$$

So,

$$x = y - \frac{1}{4},$$

which is option D.

15.   The condition that the system has infinitely many solutions implies that the two equations are really the same equation. Multiplying the first equation by 6, the system becomes

$$6ax + 6by = 54$$
$$3x + 4y = 54$$

Hence,
$$6a = 3 \text{ and } 6b = 4.$$

So,
$$a = \frac{3}{6} \text{ and } b = \frac{4}{6}.$$

Therefore,
$$\frac{a}{b} = \frac{\left(\frac{3}{6}\right)}{\left(\frac{4}{6}\right)} = \frac{3}{6} \cdot \frac{6}{4} = \frac{3}{4}.$$

17.   The condition that the system has no solution implies that the two equations represent two distinct but parallel lines. Hence, the slopes of the lines are the same. Converting the equations into slope-intercept form, we have

$$kx - 2y = 5 \quad \Longrightarrow \quad kx - 5 = 2y \quad \Longrightarrow \quad \frac{k}{2}x - \frac{5}{2} = y \quad \Longrightarrow \quad y = \frac{k}{2}x - \frac{5}{2}$$

and

$$3x - 4y = 8 \quad \Longrightarrow \quad 3x - 8 = 4y \quad \Longrightarrow \quad \frac{3}{4}x - \frac{8}{4} = y \quad \Longrightarrow \quad y = \frac{3}{4}x - 2.$$

(Note that the equations have distinct y-intercepts.) So,

$$\frac{k}{2} = \frac{3}{4}.$$

Thus,
$$k = \frac{6}{4} = \frac{3}{2}$$

is the value of k that will render the system of equations to have no solution.

19. Since we want the equation that represents a line parallel to the equation

$$y = -4x + 4,$$

we need to find the equation that represents a line with the same slope, viz., $m = -4$. Converting equation C into slope-intercept form, we have

$$8x + 2y = 6 \quad \Longrightarrow \quad 2y = -8x + 6 \quad \Longrightarrow \quad y = -4x + 3.$$

Hence, equation C has the slope $m = -4$. So, option C is the correct answer.

21. Converting the second equation into slope-intercept form, we have

$$2y - 2x = 6 \quad \Longrightarrow \quad 2y = 2x + 6 \quad \Longrightarrow \quad y = x + 3.$$

Hence, the slope of both lines is $m = 1$. However, the y-intercepts of the two lines are not the same. Therefore, we have two distinct but parallel lines, viz., option A.

22. Converting the second equation into slope-intercept form, we have

$$3x - 4y = 10 \quad \Longrightarrow \quad -4y = -3x + 10 \quad \Longrightarrow \quad y = \frac{3}{4}x - \frac{5}{2}.$$

Hence, the slopes of both lines are neither the same nor negative reciprocals of one another. So, options A, B, and C cannot be true. As extra confirmation that option D is correct, note that by substituting $x = -30$ and $y = -25$ we get the true equations

$$-25 = -30 + 5$$
$$3(-30) - 4(-25) = 10.$$

23. Converting the second equation into slope-intercept form, we have

$$x + 2y = 16 \quad \Longrightarrow \quad 2y = -x + 16 \quad \Longrightarrow \quad y = -\frac{1}{2}x + 8.$$

Hence, the slopes of both lines are negative reciprocals of one another. So, option C is correct.

5. Complex Numbers

In essence, for the problems involving complex numbers, there are two definitions and three operations to know about.

Firstly, like the Greek letter π, which has a special value in mathematics, the letter i also has a particular value. That is, $i^2 = 1$ or $i = \sqrt{-1}$. Secondly, a "complex number" is a number of the form $a + bi$, where both a and b are real numbers, and $i = \sqrt{-1}$. In the expression $a + bi$, the number a is called the "real part" and the number bi is called the "imaginary part." For example, in the complex number

$$3 + 4i,$$

the number 3 is the real part and $4i$ is the imaginary part.

Secondly, the three operations that come up are addition/subtraction, multiplication, and division.

For addition and subtraction, we add (subtract) the real parts and then add (subtract) the imaginary parts. For example,

$$(3 + 5i) + (4 - i) = 3 + 4 + 5i - i$$
$$= (3 + 4) + (5 - 1)i$$
$$= 7 + 4i$$

The easiest way to handle multiplication is by the FOIL method. Remember to apply the definition that $i^2 = 1$. For example,

$$(3 + 5i)(4 - i) = \overset{F}{\overbrace{3 \cdot 4}} - \overset{O}{\overbrace{3 \cdot i}} + \overset{I}{\overbrace{4 \cdot 5i}} - \overset{L}{\overbrace{5i^2}}$$
$$= 12 - 3i + 20i - 5i^2$$
$$= 12 + 17i - 5(-1)$$
$$= 12 + 17i + 5$$
$$= 12 + 5 + 17i$$
$$= 17 + 17i$$

Finally we come to the division of two complex numbers, for example $\frac{3+5i}{4-i}$. Crucial to solving such problems is the concept of the "conjugate" of a complex number. The conjugate of $4 - i$ is $4 + i$. Likewise, the conjugate of $4 + i$ is $4 - i$. In general, $a + bi$ and $a - bi$ are conjugates of one another. Now to solve $\frac{3+5i}{4-i}$, proceed as follows:

$$\frac{3 + 5i}{4 - i} = \frac{3 + 5i}{4 - i} \cdot \frac{4 + i}{4 + i}$$

43

$$= \frac{(3+5i)}{(4-i)} \cdot \frac{(4+i)}{(4+i)}$$

$$= \frac{3 \cdot 4 + 3i + 4 \cdot 5i + 5i^2}{4 \cdot 4 + 4i - 4i - i^2}$$

$$= \frac{12 + 3i + 20i + 5i^2}{16 - i^2}$$

$$= \frac{12 + 23i + 5i^2}{16 - i^2}$$

$$= \frac{12 + 23i + 5(-1)}{16 - (-1)}$$

$$= \frac{12 + 23i - 5}{16 + 1}$$

$$= \frac{7 + 23i}{17}$$

$$= \frac{7}{17} + \frac{23}{17}i$$

The College Board presents problems involving complex numbers in three formats. Here is one example of each format:

1. For $i = \sqrt{-1}$, what is the sum $(8 + 4i) + (-7 + 10i)$?

2. Which of the following complex numbers is equivalent to $\frac{5-3i}{10+4i}$? (Note: $i = \sqrt{-1}$)

A) $\frac{5}{10} - \frac{3i}{4}$

B) $\frac{5}{10} + \frac{3i}{4}$

C) $\frac{19}{58} - \frac{25i}{58}$

D) $\frac{19}{58} + \frac{25i}{58}$

3.

$$\frac{7-i}{4-3i}$$

If the expression above is rewritten in the form $a + bi$, where a and b are real numbers, what is the value of a? (Note: $i = \sqrt{-1}$)

To solve the first problem, compute the sum by adding the real parts and the imaginary parts.

$$(8 + 4i) + (-7 + 10i) = 8 - 7 + 4i + 10i$$
$$= (8 - 7) + (4 + 10)i$$
$$= 1 + 14i$$

To solve the second problem, start by multiplying the numerator and denominator by the conjugate of the denominator.

$$\frac{5-3i}{10+4i} = \frac{5-3i}{10+4i} \cdot \frac{10-4i}{10-4i}$$
$$= \frac{(5-3i)}{(10+4i)} \cdot \frac{(10-4i)}{(10-4i)}$$
$$= \frac{5 \cdot 10 - 5 \cdot 4i - 10 \cdot 3i + 3 \cdot 4i^2}{10 \cdot 10 - 10 \cdot 4i + 10 \cdot 4i - 4 \cdot 4i^2}$$
$$= \frac{50 - 20i - 30i + 12i^2}{100 - 40i + 40i - 16i^2}$$
$$= \frac{50 - 50i + 12i^2}{100 - 16i^2}$$
$$= \frac{50 - 50i + 12(-1)}{100 - 16(-1)}$$
$$= \frac{50 - 50i - 12}{100 + 16}$$
$$= \frac{38 - 50i}{116}$$
$$= \frac{38}{116} - \frac{50i}{116}$$
$$= \frac{19}{58} - \frac{25i}{58}$$

Solving the third problem is very similar to the second problem. The only difference is that the final answer only specifies the real part of the complex number.

$$\frac{7-i}{4-3i} = \frac{(7-i)}{(4-3i)} \cdot \frac{(4+3i)}{(4+3i)}$$

$$= \frac{28 + 17i - 3i^2}{16 - 9i^2}$$

$$= \frac{31 - 17i}{25}$$

$$= \frac{31}{25} - \frac{17i}{25}$$

The real part being the number $\frac{31}{25}$.

PROBLEMS

1. For $i = \sqrt{-1}$, what is the sum $(2 + 3i) + (4 + 5i)$?

2. For $i = \sqrt{-1}$, what is the sum $(5 + 2i) + (7 + 6i)$?

3. For $i = \sqrt{-1}$, what is the sum $(4 + i) + (2 + 10i)$?

4. For $i = \sqrt{-1}$, what is the sum $(6 + 4i) + (3 + i)$?

5. For $i = \sqrt{-1}$, what is the sum $(7 + 5i) + (-3 + 10i)$?

6. For $i = \sqrt{-1}$, what is the sum $(8 + 3i) + (-6 + i)$?

7. For $i = \sqrt{-1}$, what is the sum $(6 + 4i) + (-7 + 9i)$?

8. For $i = \sqrt{-1}$, what is the sum $(-4 + 4i) + (3 + 7i)$?

9. For $i = \sqrt{-1}$, what is the sum $(-5 + 4i) + (7 - 8i)$?

10. For $i = \sqrt{-1}$, what is the sum $(-3 + 2i) + (4 - 10i)$?

11. Which of the following complex numbers is equivalent to $\frac{1-3i}{6+2i}$? (Note: $i = \sqrt{-1}$)

A) $\frac{i}{2}$

B) $-\frac{i}{2}$

C) $\frac{1}{6} - \frac{3i}{2}$

D) $\frac{1}{6} + \frac{3i}{2}$

12. Which of the following complex numbers is equivalent to $\frac{5-7i}{10+4i}$? (Note: $i = \sqrt{-1}$)

A) $\frac{5}{10} + \frac{7i}{4}$

B) $\frac{5}{10} - \frac{7i}{4}$

C) $\frac{11}{58} - \frac{45i}{58}$

D) $\frac{11}{58} + \frac{45i}{58}$

13. Which of the following complex numbers is equivalent to $\frac{9-5i}{6+8i}$? (Note: $i = \sqrt{-1}$)

A) $\frac{7}{50} - \frac{51i}{50}$

B) $\frac{7}{50} + \frac{51i}{50}$

C) $\frac{9}{6} - \frac{5i}{8}$

D) $\frac{9}{6} + \frac{5i}{8}$

14. Which of the following complex numbers is equivalent to $\frac{11-3i}{8+2i}$? (Note: $i = \sqrt{-1}$)

A) $\frac{11}{8} - \frac{3i}{2}$

B) $\frac{41}{34} + \frac{3i}{4}$

C) $\frac{41}{34} - \frac{23i}{34}$

D) $\frac{11}{8} + \frac{3i}{2}$

15. Which of the following complex numbers is equivalent to $\frac{1+i}{1-i}$? (Note: $i = \sqrt{-1}$)

A) $1 + i$

B) $1 - i$

C) i

D) $-i$

16. Which of the following complex numbers is equivalent to $\frac{1-i}{1+i}$? (Note: $i = \sqrt{-1}$)

A) i

B) $-i$

C) $1 + i$

D) $1 - i$

17. Which of the following complex numbers is equivalent to $\frac{2+i}{2-i}$? (Note: $i = \sqrt{-1}$)

A) $\frac{2}{2} - i$

B) $1 - i$

C) $\frac{3}{5} - \frac{4i}{5}$

D) $\frac{3}{5} + \frac{4i}{5}$

18. Which of the following complex numbers is equivalent to $\frac{-5-2i}{4-i}$? (Note: $i = \sqrt{-1}$)

A) $\frac{-5}{4} + 2i$

B) $\frac{-5}{4} - 2i$

C) $\frac{-18}{17} + \frac{13i}{17}$

D) $\frac{-18}{17} - \frac{13i}{17}$

19. Which of the following complex numbers is equivalent to $\frac{-3-5i}{7-2i}$? (Note: $i = \sqrt{-1}$)

A) $\frac{-3}{7} + \frac{5i}{2}$

B) $\frac{3}{7} - \frac{5i}{2}$

C) $\frac{-11}{53} - \frac{41i}{53}$

D) $\frac{11}{53} + \frac{41i}{53}$

20. Which of the following complex numbers is equivalent to $\frac{-3+2i}{2-5i}$? (Note: $i = \sqrt{-1}$)

A) $\frac{-16}{29} + \frac{11i}{29}$

B) $\frac{-16}{29} - \frac{11i}{29}$

C) $\frac{-3}{2} + \frac{2i}{5}$

D) $\frac{-3}{2} - \frac{2i}{5}$

21.
$$\frac{6-i}{3-2i}$$

If the expression above is rewritten in the form $a + bi$, where a and b are real numbers, what is the value of a? (Note: $i = \sqrt{-1}$)

22.
$$\frac{5-2i}{4-3i}$$

If the expression above is rewritten in the form $a + bi$, where a and b are real numbers, what is the value of a? (Note: $i = \sqrt{-1}$)

23.
$$\frac{8+4i}{6-5i}$$

If the expression above is rewritten in the form $a + bi$, where a and b are real numbers, what is the value of a? (Note: $i = \sqrt{-1}$)

24.
$$\frac{2+3i}{4-7i}$$

If the expression above is rewritten in the form $a + bi$, where a and b are real numbers, what is the value of a? (Note: $i = \sqrt{-1}$)

25.

$$\frac{8 - 3i}{3 + 2i}$$

If the expression above is rewritten in the form $a + bi$, where a and b are real numbers, what is the value of a? (Note: $i = \sqrt{-1}$)

26.

$$\frac{2 - 5i}{4 + 3i}$$

If the expression above is rewritten in the form $a + bi$, where a and b are real numbers, what is the value of a? (Note: $i = \sqrt{-1}$)

27.

$$\frac{4 + 3i}{5 + 2i}$$

If the expression above is rewritten in the form $a + bi$, where a and b are real numbers, what is the value of a? (Note: $i = \sqrt{-1}$)

28.

$$\frac{4 + 6i}{3 + 5i}$$

If the expression above is rewritten in the form $a + bi$, where a and b are real numbers, what is the value of a? (Note: $i = \sqrt{-1}$)

29.

$$\frac{-4 - 2i}{-3 - 5i}$$

If the expression above is rewritten in the form $a + bi$, where a and b are real numbers, what is the value of a? (Note: $i = \sqrt{-1}$)

30.

$$\frac{-10 - 4i}{5 + i}$$

If the expression above is rewritten in the form $a + bi$, where a and b are real numbers, what is the value of a? (Note: $i = \sqrt{-1}$)

1. $6 + 8i$	11. B	21. $\frac{20}{13}$
2. $12 + 8i$	12. C	22. $\frac{26}{25}$
3. $6 + 11i$	13. A	23. $\frac{28}{61}$
4. $9 + 5i$	14. C	24. $-\frac{13}{65}$
5. $4 + 15i$	15. C	25. $\frac{18}{13}$
6. $2 + 4i$	16. B	26. $-\frac{7}{25}$
7. $-1 + 13i$	17. D	27. $\frac{26}{29}$
8. $-1 + 11i$	18. D	28. $\frac{21}{17}$
9. $2 - 4i$	19. C	29. $\frac{11}{17}$
10. $1 - 8i$	20. B	30. $-\frac{27}{13}$

SELECTED SOLUTIONS

1.

$$(2 + 3i) + (4 + 5i) = 2 + 4 + 3i + 5i$$
$$= (2 + 4) + (3 + 5)i$$
$$= 6 + 8i$$

11.

$$\frac{1 - 3i}{6 + 2i} = \frac{1 - 3i}{6 + 2i} \cdot \frac{6 - 2i}{6 - 2i}$$

$$= \frac{(1 - 3i)}{(6 + 2i)} \cdot \frac{(6 - 2i)}{(6 - 2i)}$$

$$= \frac{1 \cdot 6 - 1 \cdot 2i - 6 \cdot 3i + 3 \cdot 2i^2}{6 \cdot 6 - 6 \cdot 2i + 6 \cdot 2i - 2 \cdot 2i^2}$$

$$= \frac{6 - 2i - 18i + 6i^2}{36 - 12i + 12i - 4i^2}$$

$$= \frac{6 - 20i + 6i^2}{36 - 4i^2}$$

$$= \frac{6 - 20i + 6(-1)}{36 - 4(-1)}$$

52

$$= \frac{6 - 20i - 6}{36 + 4}$$

$$= \frac{-20i}{40}$$

$$= -\frac{i}{2}$$

21.

$$\frac{6 - i}{3 - 2i} = \frac{(6 - i)}{(3 - 2i)} \cdot \frac{(3 + 2i)}{(3 + 2i)}$$

$$= \frac{18 + 9i - 2i^2}{9 - 4i^2}$$

$$= \frac{20 + 9i}{13}$$

$$= \frac{20}{13} + \frac{9i}{13}$$

Hence, a, the real part, is the number $\frac{20}{13}$.

6. Geometry

The College Board presents problems involving geometry in four formats. Here is one example of each format:

1. In triangle ABC, the measure of $\angle B$ is 90°, $BC = 12$, and $AC = 15$. Triangle DEF is similar to triangle ABC, where vertices D, E, and F correspond to vertices A, B, and C, respectively, and each side of triangle DEF is $\frac{1}{3}$ the length of the corresponding side of triangle ABC. What is the value of $\sin F$?

2.

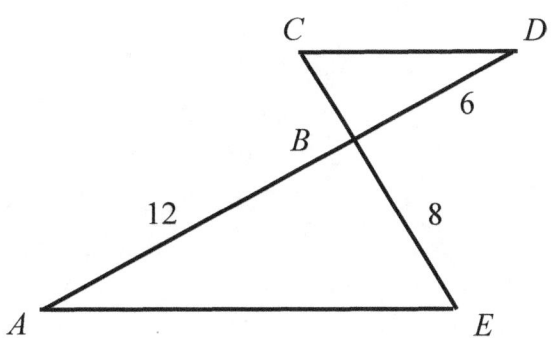

In the figure above, $\overline{AE} \parallel \overline{CD}$ and segment AD intersects segment CE at B. What is the length of segment CE ?

3.

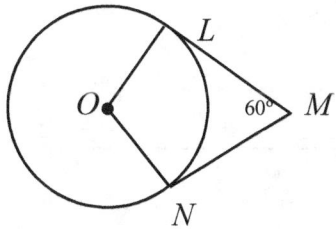

In the figure above, point O is the center of the circle, line segments LM and MN are tangent to the circle at points L and N, respectively, and the segments intersect at point M as shown. If the circumference of the circle is 93, what is the length of minor arc $\overset{\frown}{LN}$?

4.

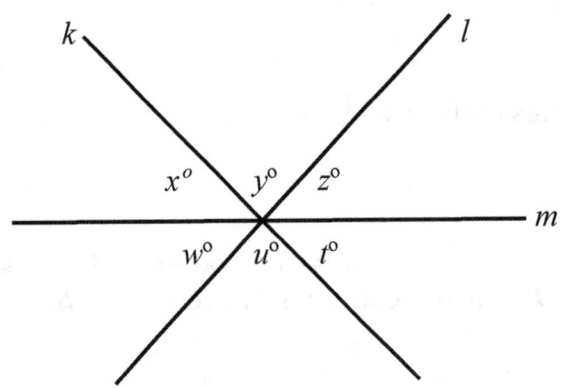

Note: Figure not drawn to scale.

In the figure above, lines k, l, and m intersect at a point. If $x + y = u + w$, which of the following must be true?

I. $x = z$
II. $y = w$
III. $x = u$

A)   I and II only
B)   I only
C)   II and III only
D)   I, II, and III

To solve the first problem, begin by drawing representative triangles.

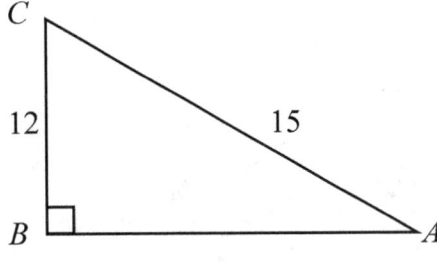

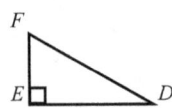

Since the sides of triangle DEF are $\frac{1}{3}$ the length of the corresponding sides of triangle ABC, we can fill in the lengths of EF and DF:

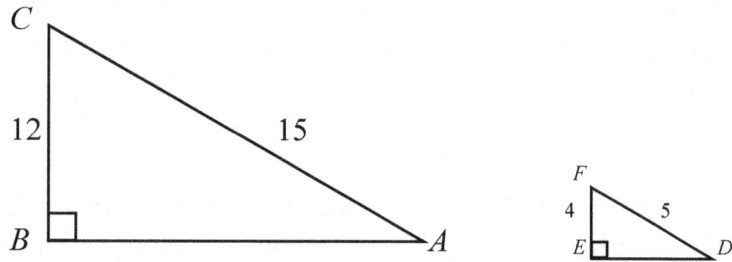

The value of $\sin F$ is equal the ratio $\frac{DE}{DF}$. Hence, we need to use the Pythagorean Theorem to find the length of DE. $DE^2 + 4^2 = 5^2$. So, $DE^2 = 25 - 16 = 9$. Thus, $DE = 3$. Therefore, $\sin F = \frac{3}{5}$.

To solve the second problem, note that $\overline{AE} \parallel \overline{CD}$. Hence, $\angle C \cong \angle E$ and $\angle D \cong \angle A$. Furthermore, vertical angles $\angle CBD$ and $\angle EBA$ are also congruent. So, $\triangle CBD$ is similar to $\triangle EBA$. Thus, $\frac{CB}{6} = \frac{8}{12}$. Hence, $CB = \frac{6 \cdot 8}{12} = \frac{48}{12} = 4$. Therefore, $CE = 4 + 8 = 12$.

To solve the third problem, first note that since LM and MN are tangent to the circle at points L and N, respectively, both $\angle MLO$ and $\angle MNO$ are right angles. Hence, $\angle MLO, \angle LMN$ and $\angle MNO$ add up to $90° + 60° + 90° = 240°$. However, all the angles of quadrilateral $OLMN$ sum up to a total of $360°$. So, $\angle LON = 360° - 240° = 120°$. Thus, since 120 is one third of 360, the length of minor arc $\overset{\frown}{LN}$ is one third of the circle's circumference. Therefore, the arc's length is $\frac{93}{3} = 31$.

Solving the fourth problem, begin by noting that vertical angles y and u are congruent. Hence, $x + y = y + w$. So, $x = w$. But vertical angles w and z are congruent. Consequently, $x = z$ (option I). Given the assumptions of this problem, equal angles y and u could both be $80°$ and both x and w could be, say, $40°$. In that case, $y \neq w$ and $x \neq u$. Therefore, only option I *must* be true—choice B.

PROBLEMS

1. In triangle ABC, the measure of $\angle B$ is $90°$, $BC = 9$, and $AC = 15$. Triangle DEF is similar to triangle ABC, where vertices D, E, and F correspond to vertices A, B, and C, respectively, and each side of triangle DEF is $\frac{1}{3}$ the length of the corresponding side of triangle ABC. What is the value of $\sin F$?

2. In triangle ABC, the measure of $\angle B$ is $90°$, $BC = 15$, and $AC = 25$. Triangle DEF is similar to triangle ABC, where vertices D, E, and F correspond to vertices A, B, and C, respectively, and each side of triangle DEF is $\frac{1}{5}$ the length of the corresponding side of triangle ABC. What is the value of $\sin F$?

3. In triangle ABC, the measure of $\angle B$ is $90°$, $BC = 16$, and $AC = 20$. Triangle DEF is similar to triangle ABC, where vertices D, E, and F correspond to vertices A, B, and C, respectively, and each side of triangle DEF is $\frac{1}{4}$ the length of the corresponding side of triangle ABC. What is the value of $\sin F$?

4.

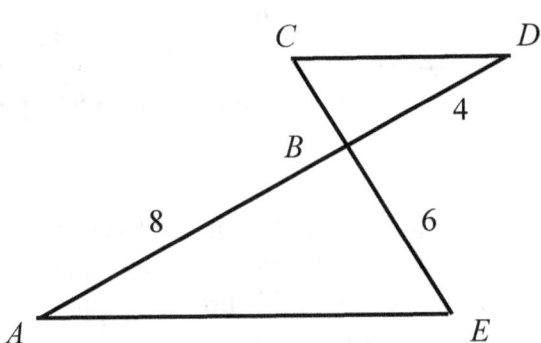

In the figure above, $\overline{AE} \parallel \overline{CD}$ and segment AD intersects segment CE at B. What is the length of segment CE ?

5.

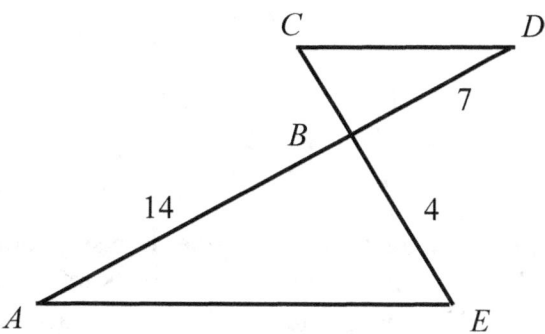

In the figure above, $\overline{AE} \parallel \overline{CD}$ and segment AD intersects segment CE at B. What is the length of segment CE ?

6.

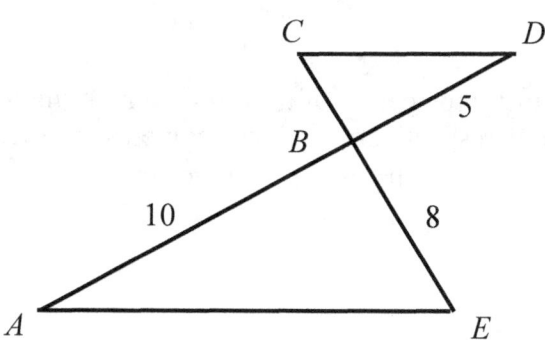

In the figure above, $\overline{AE} \parallel \overline{CD}$ and segment AD intersects segment CE at B. What is the length of segment CE ?

7.

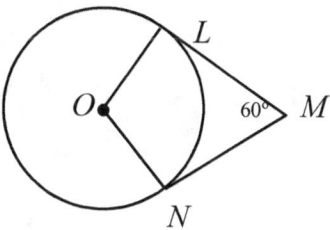

In the figure above, point O is the center of the circle, line segments LM and MN are tangent to the circle at points L and N, respectively, and the segments intersect at point M as shown. If the circumference of the circle is 99, what is the length of minor arc $\overset{\frown}{LN}$?

8.

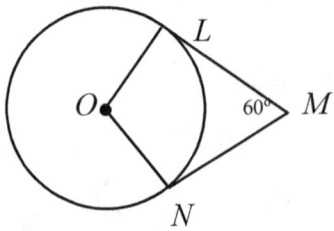

In the figure above, point O is the center of the circle, line segments LM and MN are tangent to the circle at points L and N, respectively, and the segments intersect at point M as shown. If the circumference of the circle is 45, what is the length of minor arc $\overset{\frown}{LN}$?

9.

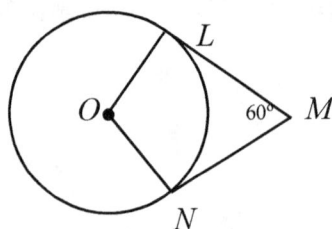

In the figure above, point O is the center of the circle, line segments LM and MN are tangent to the circle at points L and N, respectively, and the segments intersect at point M as shown. If the circumference of the circle is 96, what is the length of minor arc $\overset{\frown}{LN}$?

10.

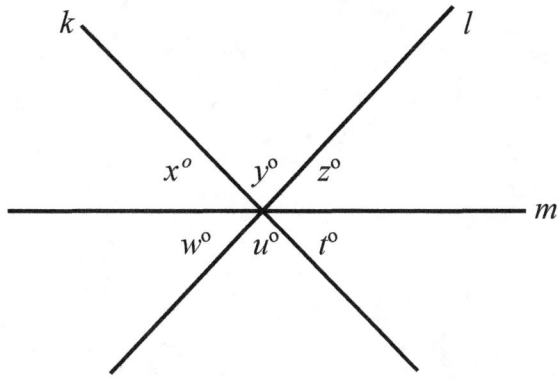

Note: Figure not drawn to scale.

In the figure above, lines k, l, and m intersect at a point. If $x + y = u + w$, which of the following must be true?

I. $w = t$
II. $y = u$
III. $w = u$

A)   I and II only
B)   I and III only
C)   II and III only
D)   I, II, and III

11.

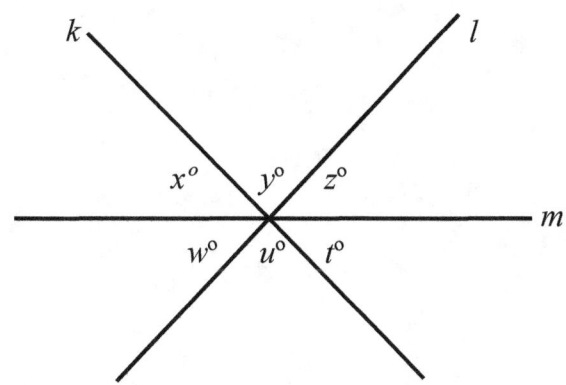

Note: Figure not drawn to scale.

In the figure above, lines k, l, and m intersect at a point. If $x + y = u + w$, which of the following must be true?

I. $y = t$
II. $z = u$
III. $y = w$

A)   I and II only
B)   I and III only
C)   II and III only
D)   None

12.

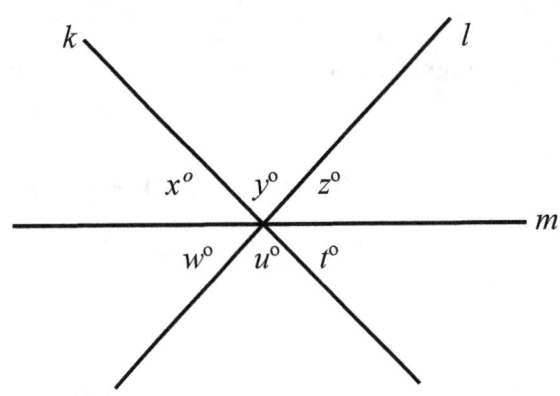

Note: Figure not drawn to scale.

In the figure above, lines k, l, and m intersect at a point. If $x + y = u + w$, which of the following must be true?

I. $x = z$
II. $y = w$
III. $z = t$

A)   I and II only
B)   I and III only
C)   II and III only
D)   I, II, and III

ANSWERS

1. $\frac{4}{5}$	5. 6	9. 32
2. $\frac{4}{5}$	6. 12	10. A
3. $\frac{3}{5}$	7. 33	11. D
4. 9	8. 15	12. B

SELECTED SOLUTIONS

1.   To solve this problem, begin by drawing representative triangles.

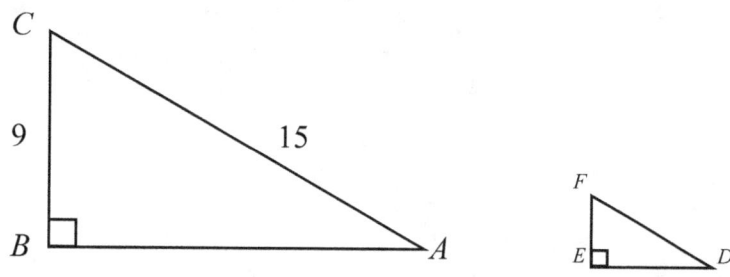

Since the sides of triangle DEF are $\frac{1}{3}$ the length of the corresponding sides of triangle ABC, we can fill in the lengths of EF and DF:

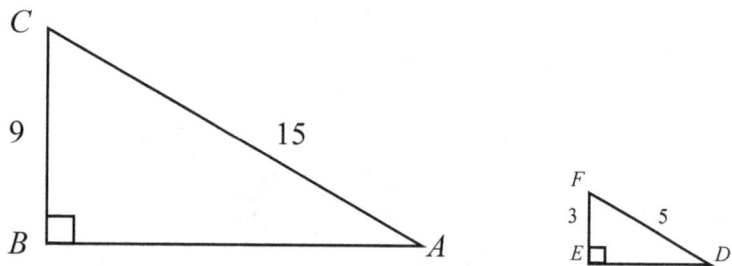

The value of $\sin F$ is equal the ratio $\frac{DE}{DF}$. Hence, we need to use the Pythagorean Theorem to find the length of DE. $DE^2 + 3^2 = 5^2$. So, $DE^2 = 25 - 9 = 16$. Thus, $DE = 4$. Therefore, $\sin F = \frac{4}{5}$.

4.   To solve this, note that $\overline{AE} \parallel \overline{CD}$. Hence, $\angle C \cong \angle E$ and $\angle D \cong \angle A$. Furthermore, vertical angles $\angle CBD$ and $\angle EBA$ are also congruent. So, $\triangle CBD$ is similar to $\triangle EBA$. Thus, $\frac{CB}{4} = \frac{6}{8}$. Hence, $CB = \frac{4 \cdot 6}{8} = \frac{24}{8} = 3$. Therefore, $CE = 3 + 6 = 9$.

63

7.   To solve this problem, first note that since *LM* and *MN* are tangent to the circle at points *L* and *N*, respectively, both $\angle MLO$ and $\angle MNO$ are right angles. Hence, $\angle MLO$, $\angle LMN$ and $\angle MNO$ add up to $90° + 60° + 90° = 240°$. However, all the angles of quadrilateral *OLMN* sum up to a total of $360°$. So, $\angle LON = 360° - 240° = 120°$. Thus, since 120 is one third of 360, the length of minor arc $\widehat{LN}$ is one third of the circle's circumference. Therefore, the arc's length is $\frac{99}{3} = 33$.

10.   In solving this problem, begin by noting that vertical angles *y* and *u* are congruent (option II). Hence, $x + y = y + w$. So, $x = w$. But vertical angles *x* and *t* are congruent. Consequently, $w = t$ (option I). Given the assumptions of this problem, equal angles *y* and *u* could both be $80°$ and both *x* and *w* could be, say, $40°$. In that case, $w \neq u$. Therefore, only options I and II *must* be true—choice A.

7. Trigonometry

In essence, for the problems involving trigonometry, there are three definitions—united by the mnemonic "SOHCAHTOA" —and one property that the College Board calls the "complementary angle relationship."

In a right triangle, there are three primary trigonometric functions, "sine," "cosine," and "tangent." These functions are defined as follows. For a given acute angle in the triangle, the "sine" of that angle is equal to the ratio of the opposite leg divided by the hypotenuse (SOH). The "cosine" of that angle is equal to the ratio of the adjacent leg divided by the hypotenuse (CAH). The "tangent" of that angle is equal to the ratio of the opposite leg divided by the adjacent leg (TOA). For example, consider the following right triangle.

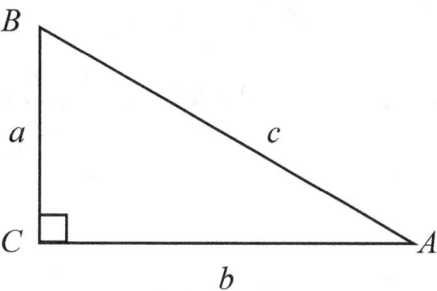

The two acute angles are labelled A and B, which are complementary, since the sum of the measures of the two angles equals $90°$. By definition, $\sin(A) = \frac{a}{c}$, $\cos(A) = \frac{b}{c}$, and $\tan(A) = \frac{a}{b}$. On the other hand, $\sin(B) = \frac{b}{c}$, $\cos(B) = \frac{a}{c}$, and $\tan(B) = \frac{b}{a}$.

Now observe that in this triangle, $\sin(A) = \frac{a}{c} = \cos(B)$ and $\sin(B) = \frac{b}{c} = \cos(A)$. Hence, in general, in a right triangle the sine of one acute angle equals the cosine of the other acute angle. Furthermore, since A and B are complementary, $B° = 90° - A°$. So, in the above triangle, $\sin(A) = \cos(90 - A)$ and $\sin(B) = \cos(90 - B)$. The generalization of these equations to all right triangles is called the "complementary angle relationship."

The College Board presents problems involving trigonometry in four formats. Here is one example of each format:

1.

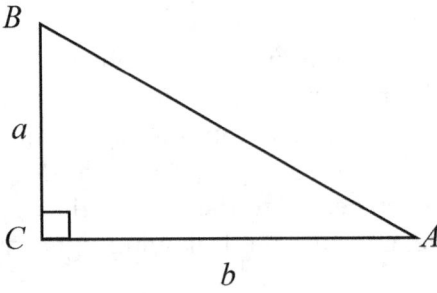

Given the right triangle ABC above, which of the following is equal to $\frac{b}{a}$?

A) $\sin A$

B) $\sin B$

C) $\tan A$

D) $\tan B$

2.

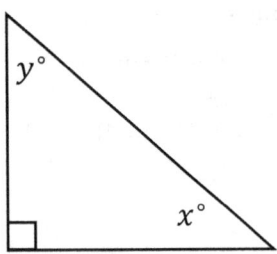

In the triangle above, the sine of $x°$ is 0.7. What is the cosine of $y°$?

3. In a right triangle, one angle measures $x°$, where $\sin x° = \frac{3}{5}$. What is $\cos(90° - x°)$?

4.

Note: Figures not drawn to scale.

The angles shown above are acute and $\sin(a°) = \cos(b°)$. If $a = 2k - 20$ and $b = 8k - 11$, what is the value of k ?

A) 4.5

B) 5.5

C) 12.1

D) 21.1

To solve the first problem, apply "SOHCAHTOA." Since $\frac{b}{a}$ is a ratio of the right triangle's two legs, the correct answer has to involve the tangent function. Hence, the correct answer can only be C or D. However, only answer D has the correct tangent function, since relative to angle B, the ratio of the opposite leg divided by the adjacent leg does indeed equal $\frac{b}{a}$.

The second problem is a straightforward application of the complementary angle relationship. Angles x and y are the two complementary angles of the given right triangle. Hence, the sine of $x°$ equals the cosine of $y°$. So, the cosine of $y°$ has to equal 0.7.

The third problem is a more indirect application of the complementary angle relationship. The angle equal to $90° - x°$ *is* the complement of angle measuring $x°$. Hence, $\sin x° = \cos(90° - x°)$. So, $\cos(90° - x°)$ has to equal $\frac{3}{5}$.

The fourth problem is the trickiest by far. We are given two acute angles consisting of $a°$ and $b°$ such that $\sin(a°) = \cos(b°)$. By the complementary angle relationship, $\sin(a°) = \cos(90° - a°)$. Hence, $b° = 90° - a°$, by substitution. So, since $a = 2k - 20$ and $b = 8k - 11$, it follows that $8k - 11 = 90 - (2k - 20)$. Thus, $8k - 11 = 90 - 2k + 20$, i.e., $10k = 121$. Therefore, $k = \frac{121}{10} = 12.1$, answer C.

67

1.

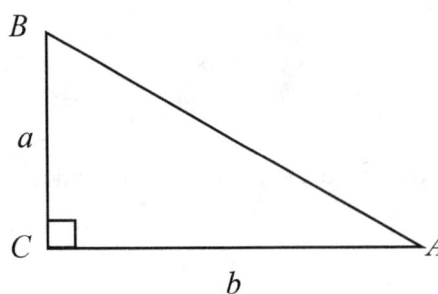

Given the right triangle ABC above, which of the following is equal to $\frac{a}{b}$?

A) $\sin A$

B) $\sin B$

C) $\tan A$

D) $\tan B$

2.

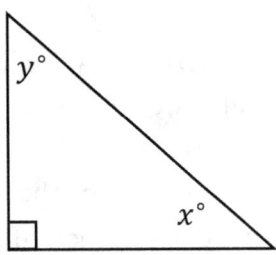

In the triangle above, the sine of $x°$ is 0.8. What is the cosine of $y°$?

3.

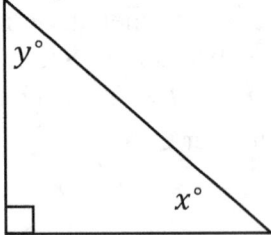

In the triangle above, the cosine of $x°$ is 0.6. What is the sine of $y°$?

4.

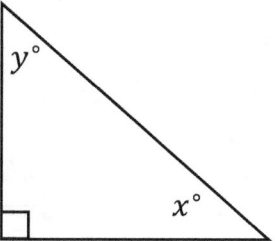

In the triangle above, the sine of $x°$ is 0.4. What is the cosine of $90° − x°$?

5. In a right triangle, one angle measures $x°$, where $\sin x° = \frac{1}{2}$. What is $\cos(90° − x°)$?

6. In a right triangle, one angle measures $x°$, where $\sin x° = \frac{4}{5}$. What is $\cos(90° − x°)$?

7.

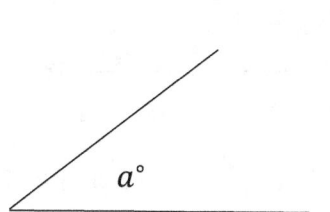

Note: Figures not drawn to scale.

The angles shown above are acute and $\sin(a°) = \cos(b°)$. If $a = 2k − 20$ and $b = 8k − 15$, what is the value of k ?

A) 3.5
B) 4.5
C) 12.5
D) 21.5

8.

Note: Figures not drawn to scale.

The angles shown above are acute and $\sin(a°) = \cos(b°)$. If $a = 6k - 18$ and $b = 4k - 19$, what is the value of k ?

A) 12.7
B) 21.7
C) 3.5
D) 4.5

9.

Note: Figures not drawn to scale.

The angles shown above are acute and $\sin(a°) = \cos(b°)$. If $a = 4k - 16$ and $b = 6k - 17$, what is the value of k ?

A) 5.5
B) 4.5
C) 12.3
D) 21.3

70

10.

Note: Figures not drawn to scale.

The angles shown above are acute and $\sin(a°) = \cos(b°)$. If $a = 5k - 22$ and $b = 5k - 12$, what is the value of k ?

A) 3.5
B) 4.5
C) 12.4
D) 21.4

1. C	4. 0.4	7. C
2. 0.8	5. $\frac{1}{2}$	8. A
3. 0.6	6. $\frac{4}{5}$	9. C
	10. C	

SELECTED SOLUTIONS

1.   To solve this problem, apply "SOHCAHTOA." Since $\frac{b}{a}$ is a ratio of the right triangle's two legs, the correct answer has to involve the tangent function. Hence, the correct answer can only be C or D. However, only answer C has the correct tangent function, since relative to angle A, the ratio of the opposite leg divided by the adjacent leg does indeed equal $\frac{b}{a}$.

2.   This problem is a straightforward application of the complementary angle relationship. Angles x and y are the two complementary angles of the given right triangle. Hence, the sine of $x°$ equals the cosine of $y°$. So, the cosine of $y°$ has to equal 0.8.

5.   This problem is a somewhat indirect application of the complementary angle relationship. The angle equal to $90° - x°$ *is* the complement of angle measuring $x°$. Hence, $\sin x° = \cos(90° - x°)$. So, $\cos(90° - x°)$ has to equal $\frac{1}{2}$.

7.   This problem is tricky. We are given two acute angles consisting of $a°$ and $b°$ such that $\sin(a°) = \cos(b°)$. By the complementary angle relationship, $\sin(a°) = \cos(90° - a°)$. Hence, $b° = 90° - a°$, by substitution. So, since $a = 2k - 20$ and $b = 8k - 15$, it follows that $8k - 15 = 90 - (2k - 20)$. Thus, $8k - 15 = 90 - 2k + 20$, i.e., $10k = 125$. Therefore, $k = \frac{125}{10} = 12.5$, answer C.

8. Randomized Problem Set 1

1. What is the sum of all values of m that satisfy $3m^2 - 12m + 3 = 0$?

2.

Note: Figures not drawn to scale.

The angles shown above are acute and $\sin(a°) = \cos(b°)$. If $a = 2k - 20$ and $b = 8k - 15$, what is the value of k ?

A) 3.5
B) 4.5
C) 12.5
D) 21.5

3.

$$(x + 4)^2 - 9 = 0$$

What is a value of x that satisfies the equation above?

4.

$$kx - 2y = 5$$
$$3x - 4y = 8$$

In the system of equations above, k is a constant and x and y are variables. For what value of k will the system of equations have no solution?

5. If $f(x) = -3x + 6$, what is $f(-2x)$ equal to?

6.

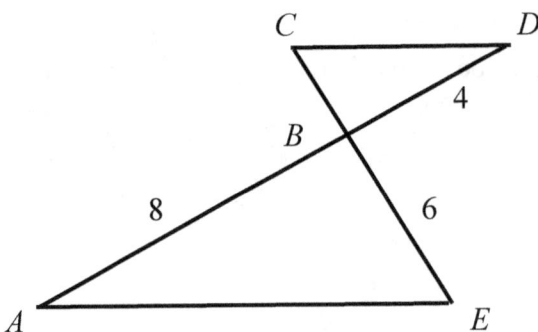

In the figure above, $\overline{AE} \parallel \overline{CD}$ and segment AD intersects segment CE at B. What is the length of segment CE ?

7. If $a = 5\sqrt{2}$ and $3a = \sqrt{2x}$, what is the value of x ?

8.

$$ax + by = 9$$
$$3x + 4y = 54$$

In the system of equations above, a and b are constants. If the system has infinitely many solutions, what is the value of $\frac{a}{b}$?

9.

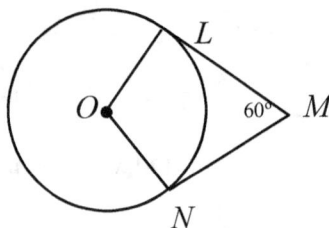

In the figure above, point O is the center of the circle, line segments LM and MN are tangent to the circle at points L and N, respectively, and the segments intersect at point M as shown. If the circumference of the circle is 99, what is the length of minor arc $\widehat{LN}$?

10.

$$f(x) = \frac{3}{2}x + b$$

In the function above, b is a constant. If $f(4) = 6$, what is the value of $f(-2)$?

11.

$$\sqrt{x - a} = x - 4$$

If $a = 2$, what is the solution set of the equation above?

12.

$$3x - 4y = -11$$
$$4x - 3y = 4$$

If (x, y) is a solution to the system of equations above, what is the value of $x - y$?

A) -15

B) -7

C) -1

D) 7

13.

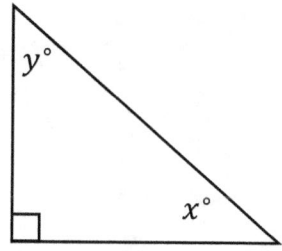

In the triangle above, the sine of $x°$ is 0.8. What is the cosine of $y°$?

14. If $x > 0$ and $5x^2 + 4x - 1 = 0$, what is the value of x?

15. Which of the following equations represents a line that is parallel to the line with equation $y = -4x + 4$?

A) $6x + 4y = 15$
B) $4x - y = 7$
C) $8x + 2y = 6$
D) $x + 2y = 1$

16. What are the solutions to the equation
$$2x^2 - 72 = 0 ?$$

17.
$$y = x - 3$$
$$2y - 2x = 6$$

The system of equations above consists of two equations, and the graph of each equation in the xy-plane is a line. Which of the following statements is true about these two lines?

A) The lines are parallel.

B) The lines are the same.

C) The lines are perpendicular.

D) The lines intersect at $(-3, 6)$.

18.

$$x^2 - \frac{k}{2}x = 2p$$

In the quadratic equation above, k and p are constants. What are the solutions for x ?

A) $x = \frac{k}{4} \pm \frac{\sqrt{k^2+2p}}{4}$

B) $x = \frac{k}{2} \pm \frac{\sqrt{k^2+32p}}{4}$

C) $x = \frac{k}{4} \pm \frac{\sqrt{k^2+2p}}{8}$

D) $x = \frac{k}{4} \pm \frac{\sqrt{k^2+32p}}{4}$

19. For $i = \sqrt{-1}$, what is the sum $(2 + 3i) + (4 + 5i)$?

20.

$$3x^2 + 6x - 9 = 0$$

If r and s are two solutions of the equation above and $r > s$, what is the value of $r - s$?

21.

$$\frac{6 - i}{3 - 2i}$$

If the expression above is rewritten in the form $a + bi$, where a and b are real numbers, what is the value of a? (Note: $i = \sqrt{-1}$)

22. In triangle ABC, the measure of $\angle B$ is $90°$, $BC = 9$, and $AC = 15$. Triangle DEF is similar to triangle ABC, where vertices D, E, and F correspond to vertices A, B, and C, respectively, and each side of triangle DEF is $\frac{1}{3}$ the length of the corresponding side of triangle ABC. What is the value of $\sin F$?

23. What are the solutions to $2x^2 + 8x + 2 = 0$?

24.

$$2x + b = 4x - 6$$
$$2y + c = 4y - 6$$

In the equations above, b and c are constants. If b is c minus $\frac{1}{2}$, which of the following is true?

A) x is y plus $\frac{1}{4}$.

B) x is y minus $\frac{1}{2}$.

C) x is y minus 1.

D) x is y minus $\frac{1}{4}$.

25.

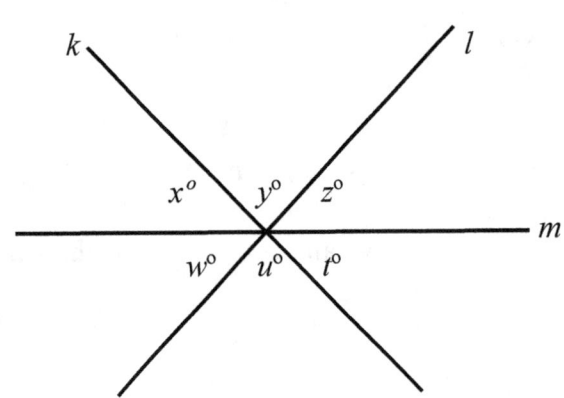

Note: Figure not drawn to scale.

In the figure above, lines k, l, and m intersect at a point. If $x + y = u + w$, which of the following must be true?

I. $w = t$
II. $y = u$
III. $w = u$

A)   I and II only
B)   I and III only
C)   II and III only
D)   I, II, and III

78

26. If $\frac{x-1}{3} = k$ and $k = 3$, what is the value of x ?

27. Which of the following complex numbers is equivalent to $\frac{1-3i}{6+2i}$? (Note: $i = \sqrt{-1}$)

A) $\frac{i}{2}$

B) $-\frac{i}{2}$

C) $\frac{1}{6} - \frac{3i}{2}$

D) $\frac{1}{6} + \frac{3i}{2}$

28.

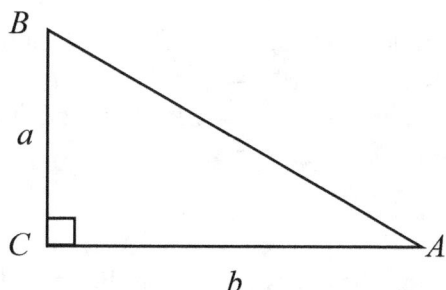

Given the right triangle ABC above, which of the following is equal to $\frac{a}{b}$?

A) $\sin A$

B) $\sin B$

C) $\tan A$

D) $\tan B$

29. If $g(x) = 3x + 2$ and $f(x) = g(x) + 5$, what is $f(2)$?

30. In a right triangle, one angle measures $x°$, where $\sin x° = \frac{1}{2}$. What is $\cos(90° - x°)$?

1. 4	7. 225	13. 0.8	19. $6 + 8i$	25. A
2. C	8. $\frac{3}{4}$	14. $\frac{1}{5}$	20. 4	26. 10
3. -1 or -7	9. 33	15. C	21. $\frac{20}{13}$	27. B
4. $\frac{3}{2}$	10. -3	16. $-6, 6$	22. $\frac{4}{5}$	28. C
5. $6x + 6$	11. $\{6\}$	17. A	23. $-2 + \sqrt{3}$, $-2 - \sqrt{3}$	29. 13
6. 9	12. C	18. D	24. D	30. $\frac{1}{2}$

SELECTED SOLUTIONS

2.   We are given two acute angles consisting of $a°$ and $b°$ such that $\sin(a°) = \cos(b°)$. By the complementary angle relationship, $\sin(a°) = \cos(90° - a°)$. Hence, $b° = 90° - a°$, by substitution. So, $a° + b° = 90°$. Thus, since $a = 2k - 20$ and $b = 8k - 15$, it follows that $(2k - 20) + (8k - 15) = 90$. Hence, $10k - 35 = 90$, i.e., $10k = 125$. Therefore, $k = \frac{125}{10} = 12.5$, answer C.

4.   The condition that the system has no solution implies that the two equations represent two distinct but parallel lines. Hence, the slopes of the lines are the same. Converting the equations into slope-intercept form, we have

$$kx - 2y = 5 \implies kx - 5 = 2y \implies \frac{k}{2}x - \frac{5}{2} = y \implies y = \frac{k}{2}x - \frac{5}{2}$$

and

$$3x - 4y = 8 \implies 3x - 8 = 4y \implies \frac{3}{4}x - \frac{8}{4} = y \implies y = \frac{3}{4}x - 2.$$

(Note that the equations have distinct y-intercepts.) So,

$$\frac{k}{2} = \frac{3}{4}.$$

Thus,

$$k = \frac{6}{4} = \frac{3}{2}$$

is the value of k that will render the system of equations to have no solution.

11.   This problem has three steps: 1. Substitute the given value of a into the equation; 2. Solve for x; 3. Test the solutions in the original equations to check for extraneous solutions. Following this procedure, we have:

$$\sqrt{x-2} = x - 4$$

$$\left(\sqrt{x-2}\right)^2 = (x-4)^2 \implies x - 2 = x^2 - 8x + 16 \implies 0 = x^2 - 9x + 18$$

$$\implies 0 = (x-6)(x-3) \implies x = 6 \text{ and } x = 3$$

Testing the solutions, we have

$$\sqrt{6-2} \stackrel{?}{=} 6 - 4$$

$$\sqrt{4} \stackrel{\checkmark}{=} 2$$

However,

$$\sqrt{3-2} \stackrel{?}{=} 3 - 4$$

$$\sqrt{1} \neq -1$$

Hence, the solution set must omit 3.

18.   Before using the quadratic formula, we need to arrange the terms of the equation in the standard order, in order to correctly identify the constants a, b, and c. Hence, we have

$$x^2 - \frac{k}{2}x - 2p = 0 \, .$$

We will make the problem easier to solve if we eliminate the fraction in the linear term. We can accomplish this by multiplying each term by 2. So, we will have

$$2x^2 - kx - 4p = 0 \, .$$

Now, using the quadratic formula, with $a = 2$, $b = -k$, and $c = -4p$, we have

$$x = \frac{-(-k) \pm \sqrt{(-k)^2 - 4(2)(-4p)}}{2(2)} = \frac{k \pm \sqrt{k^2 + 32p}}{4} = \frac{k}{4} \pm \frac{\sqrt{k^2 + 32p}}{4} \, .$$

Therefore, the answer is option D.

21.

$$\frac{6-i}{3-2i} = \frac{(6-i)}{(3-2i)} \cdot \frac{(3+2i)}{(3+2i)}$$

$$= \frac{18 + 9i - 2i^2}{9 - 4i^2}$$

$$= \frac{20 + 9i}{13}$$

$$= \frac{20}{13} + \frac{9i}{13}$$

Hence, a, the real part, is the number $\frac{20}{13}$.

82

22.   To solve this problem, begin by drawing representative triangles.

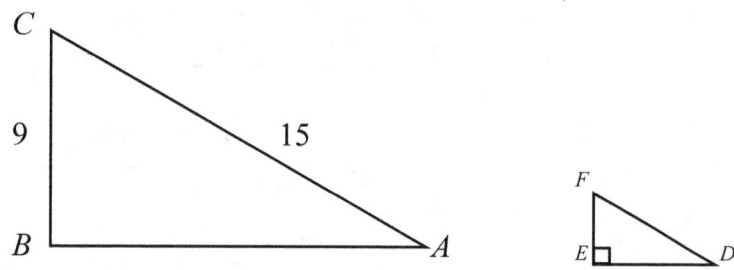

Since the sides of triangle DEF are $\frac{1}{3}$ the length of the corresponding sides of triangle ABC, we can fill in the lengths of EF and DF:

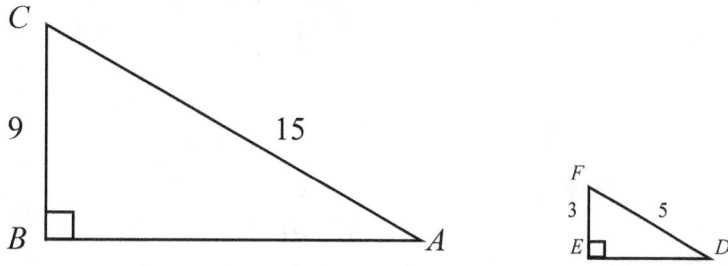

The value of $\sin F$ is equal the ratio $\frac{DE}{DF}$. Hence, we need to use the Pythagorean Theorem to find the length of DE. $DE^2 + 3^2 = 5^2$. So, $DE^2 = 25 - 9 = 16$. Thus, $DE = 4$. Therefore, $\sin F = \frac{4}{5}$.

24.   First note that the problem states $= c - \frac{1}{2}$. Hence, by substitution into the first equation,
$$2x + \left(c - \frac{1}{2}\right) = 4x - 6.$$
So,
$$c = 2x - 6 + \frac{1}{2} = 2x - \frac{12}{2} + \frac{1}{2} = 2x - \frac{11}{2}.$$
Thus, by substitution into the second equation,
$$2y + \left(2x - \frac{11}{2}\right) = 4y - 6.$$
Hence,
$$2x = 2y - 6 + \frac{11}{2} = 2y - \frac{12}{2} + \frac{11}{2} = 2y - \frac{1}{2}.$$
So,
$$x = y - \frac{1}{4},$$
which is option D.

25.

83

$$\frac{1-3i}{6+2i} = \frac{1-3i}{6+2i} \cdot \frac{6-2i}{6-2i}$$

$$= \frac{(1-3i)}{(6+2i)} \cdot \frac{(6-2i)}{(6-2i)}$$

$$= \frac{1 \cdot 6 - 1 \cdot 2i - 6 \cdot 3i + 3 \cdot 2i^2}{6 \cdot 6 - 6 \cdot 2i + 6 \cdot 2i - 2 \cdot 2i^2}$$

$$= \frac{6 - 2i - 18i + 6i^2}{36 - 12i + 12i - 4i^2}$$

$$= \frac{6 - 20i + 6i^2}{36 - 4i^2}$$

$$= \frac{6 - 20i + 6(-1)}{36 - 4(-1)}$$

$$= \frac{6 - 20i - 6}{36 + 4}$$

$$= \frac{-20i}{40}$$

$$= -\frac{i}{2},$$

which is option B.

9. Randomized Problem Set 2

1.

$$5x^2 + 7x - 6 = 0$$

If r and s are two solutions of the equation above and $r > s$, what is the value of $r - s$?

2.

$$y = x + 5$$
$$3x - 4y = 10$$

The system of equations above consists of two equations, and the graph of each equation in the xy-plane is a line. Which of the following statements is true about these two lines?

A) The lines are parallel.

B) The lines are the same.

C) The lines are perpendicular.

D) The lines intersect at $(-30, -25)$.

3.

$$\sqrt{x - a} = x - 4$$

If $a = 4$, what is the solution set of the equation above?

4.

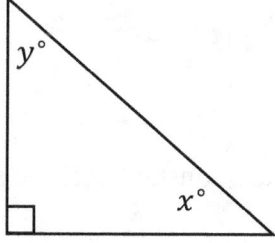

In the triangle above, the cosine of $x°$ is 0.6. What is the sine of $y°$?

5.

$$\frac{5 - 2i}{4 - 3i}$$

If the expression above is rewritten in the form $a + bi$, where a and b are real numbers, what is the value of a? (Note: $i = \sqrt{-1}$)

6. If $a = 7\sqrt{2}$ and $3a = \sqrt{2x}$, what is the value of x ?

7.

Note: Figures not drawn to scale.

The angles shown above are acute and $\sin(a°) = \cos(b°)$. If $a = 6k - 18$ and $b = 4k - 19$, what is the value of k ?

A) 12.7
B) 21.7
C) 3.5
D) 4.5

8. What are the solutions to $3x^2 + 12x + 6 = 0$?

9.

$$ax + by = 11$$
$$2x + 6y = 77$$

In the system of equations above, a and b are constants. If the system has infinitely many solutions, what is the value of $\frac{a}{b}$?

10.

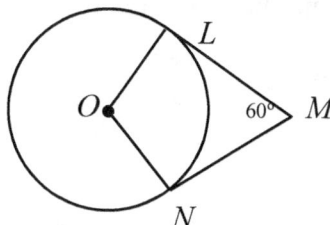

In the figure above, point O is the center of the circle, line segments LM and MN are tangent to the circle at points L and N, respectively, and the segments intersect at point M as shown. If the circumference of the circle is 45, what is the length of minor arc $\overset{\frown}{LN}$?

11. If $x > 0$ and $3x^2 + 5x - 2 = 0$, what is the value of x?

12.

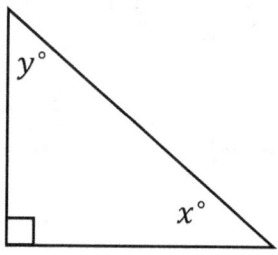

In the triangle above, the sine of $x°$ is 0.4. What is the cosine of $90° - x°$?

13.

$$(x + 2)^2 - 9 = 0$$

What is a value of x that satisfies the equation above?

14.

$$kx - 5y = 3$$
$$6x - 7y = 6$$

In the system of equations above, k is a constant and x and y are variables. For what value of k will the system of equations have no solution?

15. Which of the following complex numbers is equivalent to $\frac{5-7i}{10+4i}$? (Note: $i = \sqrt{-1}$)

A) $\frac{5}{10} + \frac{7i}{4}$

B) $\frac{5}{10} - \frac{7i}{4}$

C) $\frac{11}{58} - \frac{45i}{58}$

D) $\frac{11}{58} + \frac{45i}{58}$

16. In a right triangle, one angle measures $x°$, where $\sin x° = \frac{4}{5}$. What is $\cos(90° - x°)$?

17. What are the solutions to the equation
$$2x^2 - 32 = 0 ?$$

18. If $f(x) = -2x + 7$, what is $f(-4x)$ equal to?

19. Which of the following equations represents a line that is parallel to the line with equation $y = 2x + 3$?

A) $6x + 4y = 3$
B) $8x - 4y = 7$
C) $8x + 2y = 7$
D) $x + 6y = 10$

20.

$$x^2 - \frac{k}{4}x = 4p$$

In the quadratic equation above, k and p are constants. What are the solutions for x ?

A) $x = \frac{k}{4} \pm \frac{\sqrt{k^2+4p}}{4}$

B) $x = \frac{k}{2} \pm \frac{\sqrt{k^2+4p}}{4}$

C) $x = \frac{k}{8} \pm \frac{\sqrt{k^2+256p}}{8}$

D) $x = \frac{k}{4} \pm \frac{\sqrt{k^2+256p}}{4}$

21.

$$f(x) = \frac{5}{2}x + b$$

In the function above, b is a constant. If $f(6) = 8$, what is the value of $f(-4)$?

22. For $i = \sqrt{-1}$, what is the sum $(5 + 2i) + (7 + 6i)$?

23.

$$2x + 3y = 16$$
$$3x - 2y = -2$$

If (x, y) is a solution to the system of equations above, what is the value of $x - y$?

A) 14

B) −18

C) 0

D) −2

24. In triangle ABC, the measure of $\angle B$ is $90°$, $BC = 15$, and $AC = 25$. Triangle DEF is similar to triangle ABC, where vertices D, E, and F correspond to vertices A, B, and C, respectively, and each side of triangle DEF is $\frac{1}{5}$ the length of the corresponding side of triangle ABC. What is the value of $\sin F$?

25.

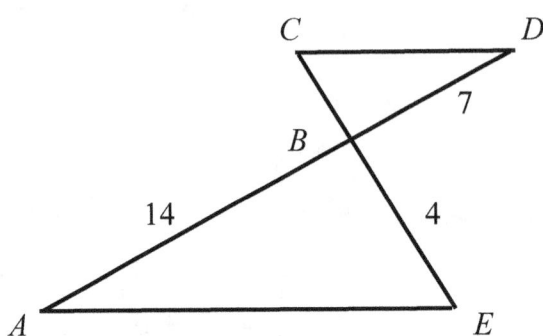

In the figure above, $\overline{AE} \parallel \overline{CD}$ and segment AD intersects segment CE at B. What is the length of segment CE ?

26. What is the sum of all values of m that satisfy $m^2 - 8m + 4 = 0$?

27.
$$3x + b = 5x - 7$$
$$3y + c = 5y - 7$$

In the equations above, b and c are constants. If b is c minus $\frac{1}{4}$, which of the following is true?

A) x is y minus $\frac{1}{4}$.

B) x is y plus $\frac{1}{2}$.

C) x is y minus $\frac{1}{8}$.

D) x is y minus 1.

28.

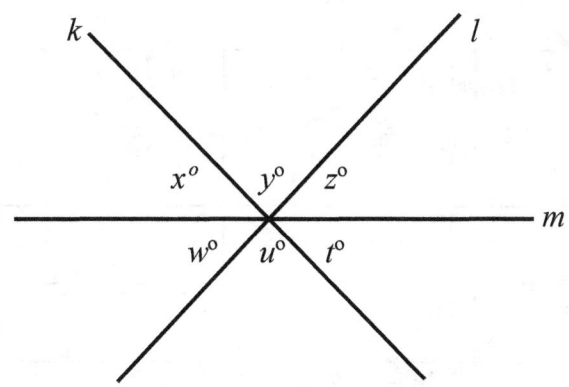

Note: Figure not drawn to scale.

In the figure above, lines k, l, and m intersect at a point. If $x + y = u + w$, which of the following must be true?

I. $y = t$
II. $z = u$
III. $y = w$

A)   I and II only
B)   I and III only
C)   II and III only
D)   None

29. If $g(x) = -x + 2$ and $f(x) = g(x) + 4$, what is $f(5)$?

30. If $\frac{x-1}{4} = k$ and $k = 5$, what is the value of x ?

ANSWERS

1. $\frac{13}{5}$	7. A	13. -5 or 1	19. B	25. 6
2. D	8. $-2 \pm \sqrt{2}$	14. $\frac{30}{7}$	20. C	26. 8
3. $\{4,5\}$	9. $\frac{1}{3}$	15. C	21. -17	27. C
4. 0.6	10. 15	16. $\frac{4}{5}$	22. $12 + 8i$	28. D
5. $\frac{26}{25}$	11. $\frac{1}{3}$	17. $-4, 4$	23. D	29. 1
6. 441	12. 0.4	18. $8x + 7$	24. $\frac{4}{5}$	30. 21

"Only he who never plays, never loses"

www.ingramcontent.com/pod-product-compliance
Lightning Source LLC
Chambersburg PA
CBHW081202180526
45170CB00006B/2190